# Programme Management
# Demystified

# Programme Management Demystified

## Managing multiple projects successfully

GEOFF REISS

London and New York

First published 1996 by E & FN Spon, an imprint of Chapman & Hall

Reprinted 1999, 2000
11 New Fetter Lane, London EC4P 4EE
29 West 35th Street, New York, NY 10001

*E & FN Spon is an imprint of the Taylor & Francis Group*

© 1996 Geoff Reiss

Printed and bound in Great Britain at TJ International Ltd, Padstow, Cornwall

*British Library Cataloguing in Publication Data*
A catalogue record for this book is available from the British Library

ISBN 0–419–21350–3 (pbk)


∞ Printed on acid-free text paper, manufactured in accordance with ANSI/NISO Z39.48-1992 and ANSI/NISO Z39.48-1984 (Permanence of Paper)

# Contents

# *Foreword*

When the idea of a book on project management was broached I went off to the quiet, peaceful village of Staithes on North Yorkshire's windy coast with a wordprocessor, a dog and a box of supermarket supplies. There I walked, ate and slept whilst the dog wrote most of *Project Management Demystified*.

The book went well. It might have helped misguide many otherwise hopeful people but they must have enjoyed reading it enough to tell their friends as sales went onward and upward. Perhaps the most pleasing fact is that colleges and universities bought copies for their students. When I was at college all textbooks had to be on the dry side of arid to make it onto the recommended reading list. Things in colleges must have improved.

It has not been a spectacularly financially-rewarding experience as the total income from the book has hardly dented my overdraft, but it has been fun. I had thought I only had one book in me but later decided there was a need for some clearing of the air about programme management – hence this book.

*Project Management Demystified* tries to cover the stuff you need to know about running one project. It tries hard to be down to earth, realistic and honest. It deals with being personally successful in project management and with running successful projects – two very different objectives. *Project Management Demystified* covers topics from defining your project through critical path and resourcing to choosing some software. It lets you sit in on a session where three people plan a small project. It moves on to special resources like money and introduces earned value analysis.

*Programme Management Demystified* starts where *Project Management Demystified* leaves off. It deals with the increasingly important topic of programme management, that is, dealing with a large number of simultaneous projects within one organization. It deals with organizational issues and the multi-project conflicts that can and do arise.

It once again remembers that your personal career prospects are at least as important as running a few hugely successful projects. It never forgets that a successful project doesn't make a successful project manager.

People take little notice of project and programme managers and planners. The better you do your job, the less notice anyone takes. An occasional toot on your own trumpet can have beneficial effects on your career. The trumpet icon indicates an opportunity for you to draw attention to your efforts in improving the company's fortunes. If handled carefully this can improve your own fortunes. If handled badly, you can easily shoot yourself in the foot.

So *Project Management Demystified* and *Programme Management Demystified* are designed to be a matched pair – one dealing with the management of single projects and the other dealing with becoming organized enough to manage many projects.

End of plug; good luck with your projects.

# *Foreplay*

I am male and I don't apologize for that. The English language gives us two singular pronouns which indicate sex and it doesn't apologize for that. I shall use both he and she, him and her from time to time, interchangeably. I don't blindly assume that all engineers are male or that all secretaries are female but I am not sure I want to become accustomed to he/she or s/he as pronouns nor can I keep finding ways to avoid any politically incorrectly written sexual presumption. And I don't apologize for that.

# *Acknowledgements*

So many people have contributed to this book I sometimes wonder what part I have played. Many of them will not want to be associated with this publication in any way but I still want to say thank you to:

My colleagues at Hydra Development Corporation Limited – Robert Fearnley, John Hopkin, Jeremy Woan and Andrew Farrell have allowed me the time to write and have kept my wordprocessor going through thick and thin. They have also filtered out the worst humour.

My colleagues at *Project Manager Today* have put up with me for years. Editor Ken Lane has quietly but unsuccessfully tried to improve my writing style and has also given me permission to use extracts from the magazine to reinforce theoretical points with case studies.

Madeleine Metcalfe, my publisher at Chapman & Hall has been a source of inspiration, royalty cheques and lunches too few to list.

Andy Wilkin, he of the pencil, paintbrush and acerbic wit, has devoted more time and energy to the cartoons in this book than his payment deserves.

David Marsh of Applied Business Technology International, genuine guru and entertainer, took the time to read through the whole draft and made many useful and sensible comments which I would not dare to ignore. Paul McBeth of Woodworth, McBeth and Young has also taken the time and trouble to read and comment sensibly on my work.

Finally I want to thank the many friends and strangers who have said nice things about *Project Management Demystified* to me at various project-management events. More than anything else, those kind people gave me the enthusiasm to get down to this book. Now you know who to blame.

# *Introduction*

## *What is programme management?*

"Up close, a mosaic is just another piece of broken glass."

Programme management means many things to many people. It is just possible that this book will make a small contribution to the crystallization of the many terms and uses for those terms that exist around this topic.

Everyone would agree that programme management is about managing a number of projects. In practice this includes most companies that are running a number of projects at the same time. This involves most organizations.

Everyone would agree that managing a programme means being able to stand back from the detailed problems and get an overview of the process as a whole. As well as looking at the individual bits of glass, you need to see the whole mosaic.

My favourite definition of programme management is:

> The co-ordinated management of a portfolio of projects which call upon the same resources.

This is not the only definition and, if you keep reading, you'll discover some more attitudes and definitions in a moment. You'll also see that programme management includes all of project management and then some.

> This is the first of many footnotes.

## *Who manages a programme?*

We are talking here of a wide range of organizations including government departments like the very useful Health, Environment and

Taxation departments. There is also the intriguing Department of Employment. Am I the only one who has bad dreams about being made redundant from the Department of Employment?

(Organizations of all kinds manage programmes.) They range from administrative organizations through computer software houses to jobbing engineering works and arms manufacturers. Such firms might have a good hold on their individual projects with existing project-management techniques or they may have decided that project-management tools do not really meet their needs.

Programme management is about the next stage of development. It involves planning each individual project so that all projects are planned and resourced. It involves planning who is going to be doing what on each project.)

## Not many people work on one project at a time

These days the trend towards programme management is so strong that few project managers are actually involved in one lonely project. The majority of projects are relatively small and run within an organization where many other projects and other endeavours are in hand.

Two of the few market sectors which are deeply involved in project management but avoid programme management are the heavy engineering and construction industries where normally each project is managed alone, separated from other projects)

Often such a project is geographically isolated, which is a nice way of saying they are stuck in the middle of some desert or in a remote valley cut off by snow for three months of the year.

When a team is set up to build a bridge, a tunnel or a zinc mine, the project manager has to fight to build his team from the specialists within the company. The brightest people are all in demand and are probably working on other projects in other corners of the world. He talks to department heads, other project managers and the personnel people and recruits from within his own company, other companies and, very occasionally, the Job Centre. This forms the key team who commit to building the bridge or road or dam for the next couple of years. Most become full-time members of the project team.

| The sex of the project manager above was selected by a random number generator |
| --- |

Many project managers gather the team around them and, in an effort to foster team spirit, commandeer the west wing of the fourth floor of head office, set up an office on site or rent some space away from headquarters. They want their team to eat, live and breath the project. They know that there are going to be enough distractions between male and females, smokers and non-smokers and over company cars. The last thing they want is the team members being borrowed to attend to some detail on their last job; to be presented to a client over a future job or to sit on the bowling green advisory committee.

Once the project team is formed, the project management becomes predominantly a matter of guiding and coercing contractors. The project manager and her team beg and plead, plan and look ahead so that the contractors will elevate the project within their in-trays. The project team generally deal with companies not people. Essentially she subcontracts the worry about resourcing to the management of the contractors.

---

This randomly selected project manager happens to be a woman.

---

Bridge builders don't get involved with individual painters and welders, just as you don't when you get a building contractor to build you a house. You might recognize the bricklayer employed by your builder but let's say that the brickie wins the pools and, having an IQ in double figures, hightails it to Bali leaving you bereft of one bricklayer. You don't go to the Job Centre, you go to the building firm as it's their job to replace the pools winner with someone else who will lay the occasional brick and demand tea at all hours.

The house or bridge builder has a strong and motivating sanction against unhelpful contractors – they can be sacked, replaced or not paid. A bridge builder will put the success of his bridge project way higher than the success of his contractors' business.

Such projects: the dams, bridges and major tunnels (but not your garage) are the rare and wonderful examples of human ingenuity and ability. The team will talk about 'them' and 'they', meaning the companies they are working with. Tunnel-building project managers say things like: 'We need more carpenters from Cumberland & Sausage.' They don't say 'Nip down to the labour exchange to get some more chippies.' These people are running a project, they are not running a programme.

In a software company managing a programme things are very different. The projects tend to be smaller and the project manager has to fight with a number of colleagues, all of whom are running their own projects and many of whom want the same design engineer or computer at just the worst moment. The information technology project manager does not normally employ contractors, but predominantly uses resources paid by the same wages department as himself.

The software engineer often has no sanctions against her colleagues whatsoever. She regards her own project as important but realizes that all the organization's projects go towards the success of the company, which is reasonably close to her heart. She must work with her fellow project managers towards the corporate goals.

Her resources are on loan to her and to her project. She might have access to the mainframe computer on Tuesday afternoons or, worse, after 7 p.m. Everyone that does something on the project does so as a favour. This favour might be structured and a part of the donor's job specification but it is still a favour. Most resources work on many other projects and many non-project related jobs. The project manager may not immediately notice when she loses someone off the project as her resources work at their own desks and terminals whatever work they are doing. They are probably only a part of her project on a part-time basis.

The priorities of the individual resources and their functional bosses are their own. Well, actually, that's not true. Their priorities look like their own but they are constantly being dragged to and fro between assorted urgent and high-priority workloads.

'Urgent' and 'high-priority' are business speak for 'unplanned'.

Sometimes 'unplanned' is OK because no one could have foreseen that a fire would damage the printing works over the weekend. Usually 'unplanned' means something that could have and should have been planned ages ago but has been lying around in the in-tray of some senior suit whilst he grapples with important issues like his paper clip supplies or his secretary. Assorted senior managers suddenly realize that because they have done nothing about 'something', that 'something' is now vital and urgent. So they pull strings, throw their weight about and steal resources from other well-planned projects. Who said life was meant to be fair? Barnum and Bailey, that's who.

Project-management Sayings #141: A lack of planning by you does not demand a panic from me.

In this environment the team is dispersed. The chances of getting the team to work in one place for a long period are vanishingly small. The project manager's only chance to achieve this laudable goal is to gather everyone for an occasional project meeting. At these meetings everyone sounds and actually may be very enthusiastic about the project, keen to be involved and willing to make their contribution. Projects tend to have that effect on people. Good project managers tend to have that effect on people.

Unfortunately, the members of the team will attend other project meetings for other projects at which they will become very enthusiastic about the project, keen to be involved and willing to make their contribution. This makes them forget about the first project and the promises they recently made.

Some thoughtful project managers will take the whole team off to a remote location (the local hotel) where the plan is explained, the team spirit cemented and communication channels opened.

(Project managers within a programme mostly deal with people – they refer to their resources by name not by skill and not by company title) You hear this sort of project manager say: 'I need Sandy for about half a day each week for the next four weeks.' If she talks about skills she will say something like this: 'I need some more programming input, can I have a few more hours of Sandy's time for the next four weeks?'

So whilst a project manager talks about getting some welders on the job, the programme manager talks about Sandy's input and Joe's time.

I have painted two extreme pictures – the large one-off project with contractors working for a main contractor and the multi-project environment where everyone works part time on a variety of projects. These two extremes are at opposite ends of a scale. Where are you on this scale? Can you see yourself, your employee and your workload in these terms?

## *Everybody's doing it*

Programme management is a real growth marketplace – it is the key growth area within the already-growing area of project management. Most project-management software houses are scrabbling to get into this marketplace and some have quite good products. A computer marketing person once said to me 'Most project-management software vendors are positioning their products in the programme-management marketplace'. I thought about this for some time before realizing what this meant and I can save you some time – it means claiming programme management features and functionality in the brochures. It may or may not mean changing the software to add some features. It certainly means changing the pamphlets.

The CCTA (the government body for this kind of thing) has published its *Introduction To Programme Management* and its fuller *Guide to Programme Management*. The details of these two publications should be in the further reading list near the back of this book. There have been seminars, colloquiums and conferences galore on this and closely-related subjects and all have been very well subscribed.

Here are some other examples of organizations that I have found on my travels which might make you realize that you are not alone in the world of programme management ills:

At any one time Thames Water plc have around 500 civil-engineering projects in hand ranging from the huge London Water Ring Main to building small local weirs and locks.

GEC Alsthom Large Machines have made a considerable investment in project and programme management, which is wise as they normally have around 700 simultaneous manufacturing projects. Wedgewood make teapots and ashtrays for the discerning buyer and had 102 product-development projects on the plan for the year when I visited them.

Hewlett Packard have around 50 software-development projects on the go involving some 120 team leaders. National Power these days manage a large number of engineering projects but employ very few productive resources as they mostly manage contractors who do the work. BT have their Martlesham Heath laboratories near Ipswich where hundreds of software-development projects are underway and National Rivers have hundreds of engineers rushing about the countryside building and improving weirs, dams, locks and fisheries.

I could rattle on for hours about such firms and what they are doing about programme management but that would be deadly boring. Instead I'll rattle on for hours about something else.

## *Projects end, programmes don't*

A friend of mine has a good approach to travel. The approach works whether the journey is by air, train or bus. It might be long or short. He claims that whatever the journey, you start, wait and then finish. For example, you get on the plane, wait and then get off. This attitude changes the journey not one bit but has a huge impact on how you feel about it. The people who rush to join long queues, are first on board and look harassed and self important are the people who don't enjoy travelling at all. My friend enjoys every moment. Similarly, given an exciting and dramatic project we can safely assume that the project will follow a predictable life cycle. It will begin, take place and end.

At some point on most projects there will be a time when the members of the project team will stand around thinking: 'There, we did it. I built that.' Simultaneously some bigwig in a suit who took virtually no interest in the project and certainly contributed nothing, will be the one saying aloud to the press: 'There, we did it. I built that.'

(Projects, like people, are born to live and die. Programmes do not. They go on for ever.) A project plan can be drawn on a piece of paper with a timescale across the top. These usually have a start date near the top left corner and an end date way down to the right.

By the way, plans drawn up in the Middle East sometimes do not follow this rule. As Arabic, Urdu (India and Pakistan), Ivrit (Israel) and Farsi (Iran) are written from right to left project plans are sometimes drawn that way.

---

You learn the strangest things in this book.

---

Now let's consider a plan for a programme. If by programme we mean one very large project like the moon landing, then we can still draw a schedule on a single bit of paper. But if we use programme to mean many internal or external projects going on throughout the life of the organization we can no longer use a single piece of paper.

To plan this we need an infinitely-long scrolling piece of paper on to which projects appear, travel across to the left and disappear. New projects are constantly being added to the right (in the future) and old, now-completed projects get deleted and fall off into the past. There is always a workload to achieve.

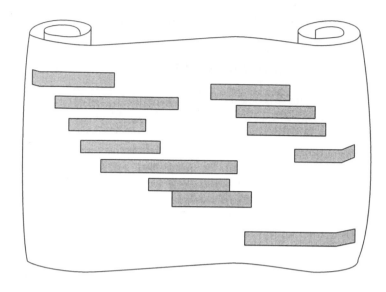

Infinitely-long scrolling pieces of paper are difficult to get these days. I scoured the Sasco catalogue and drove my local stationery store-man bonkers. I searched through catalogues of tiny brushes to clean your phone and miniature desktop aluminium dustbins to hold paperclips but to no avail. There is a shortage of infinitely-long scrolling pieces of paper.

I think that we will have to resort to some kind of electronic gadgetry to give us an everlasting plan – this sounds like a job for a computer. We'll talk about this later (p. 144).

There are people who use the term 'programme management' to mean a collection of projects each of which contributes to one specific goal and these people will be entirely happy with a normal flat piece of paper. In such an environment the programme does come to an end. We'll talk about definitions of programme management in a little while.

## Oxymorons

I am grateful to Bill Olford of Innate Software, an Englishman living in America, for the term oxymoron. An oxymoron is a phrase containing contradictory terms. Examples include a deafening silence, a near miss and military intelligence. Bill suggests that multi-project management is an oxymoron as project management is about managing one specific project and 'multi' indicates more than one.

He also distinguishes between independent and integrated. In the world of the single project the team is very independent of other projects. Members of a group sent out to a foreign land to build a dam are very independent of the other work going on within the company back home. They are dependent on the political situation in the country, the weather and the price of cheese but these are outside influences.

In a multi-project environment every project is integrated with every other project. There is little independence. There is much inter-dependence. Jobs share resources with each other and are often logically connected. For example, we could not start testing the new teddy bear until the toy testing laboratory is complete.

When we come to planning the workload we need to think again about what we expect to get from the planning process. There are clear differences between programme management and project management, especially when it comes to planning. Some of the observable trends are summarized below:

**Table 1.1** Programme planning versus project planning

| PROGRAMME MANAGEMENT | PROJECT MANAGEMENT |
| --- | --- |
| Many simultaneous projects | One project at a time |
| Personal relationship with skilled resources | Impersonal relationship with unskilled resources |
| Concentration on resources | Resources less important |
| Need to maximize utilisation of resources | Need to minimize demand for resources |
| Projects tend to be similar to each other | Projects tend to be dissimilar to each other |
| The team must ensure that the project's aim helps the organization forward | The team do not care what happens to the project after they finish their part |
| Concentration on the corporate objectives | Concentration on the project alone |
| There are loads of available tools | There are few available tools |

I'm going to expand on some of these points in a little more detail. Talk amongst yourselves for a moment. Project managers have the blissfully easy and rewarding task of concentrating on one project at a time. It might seem hard to connect the UK to France with a 26-mile tube or get a bridge built but the 'hardness' comes from size, bulk and sheer enormity. The team are able to concentrate exclusively on their project and I envy them that single-mindedness.

They say that project management is like juggling three balls – time, cost and resources – and it is true and hard to do. Programme management is like a troupe of circus performers standing in a circle, all juggling three balls simultaneously and swapping balls from time to time. Each project has its own restraints of time, cost and resources and must also be seen in terms of its effect on other projects and resources. If programme management takes place in the normal three-dimensional world then project management takes place in a flat two-dimensional world.

Programme managers have to establish and maintain teams for each project and watch for interactions between the teams, the resources and the projects themselves. In a single project there is usually a single deliverable which, one day, will be surrounded by proud project team members all saying 'I did that'. In programme management there are many deliverables some of which link into other projects. The end of a project means that one objective has been reached but the team must watch to ensure that every project is still valid and worthwhile within the moving and changing world of commerce.

On a single project there are usually resources involved somewhere along the line but very often the actual hiring and firing, guiding and checking of individual resources is subcontracted. Yes of course you need to recruit resources to form the project team, you may hire people to carry out some specific function, but increasingly the single project manager deals with other companies, each of which deals with individual resources. Sometimes the subcontractors subcontract the work to those companies who actually employ the resources that do productive work. This is a sneaky way of expanding your management team as each subcontractor contributes something to the management of the project. If you are running a one-off project, the fewer people you hire in the fewer you have to pay and fewer you have to 'let go' when the project is over.

*'Letting you go'*

Where did the phrase 'let go' come from? It is supposed to make you feel better by giving you ideas of freedom and individual choice. You have, the message infers, been let go to seek your own path, your own fortune. Actually you have been 'let go' in the same way as a mountaineer hanging from a rope. You've been given the freedom to seek the bottom of the ravine.

Therefore, on the large single project, resources tend to be involved on the project full time for a part of the life of the job. Many such people are highly mobile and can be found living in 'mobile homes' around motorway building projects or in local bedsits in the nearest town. These are people who spend their working lives leaving home on Monday morning at 6.00 a.m. and get home at 8.00 p.m. on Friday.

Resources in programmes tend to be involved part time in each project and possibly part time on the programme as a whole. Typically, specialist engineers are available to a project on Wednesday afternoons or two days per week. They are dragged from job to job and can concentrate on none.

Single projects tend to be a new unusual challenge and planners and managers alike have to burn the midnight oil figuring out how they are actually going to achieve the project. The nature of the one-off project is usually unusual. Whilst some guidance can be gleaned from previous projects, the team often have significant challenges facing them. 'How do we...?' and 'How can we...?' are common questions. The team's specialist knowledge about bridge building applied through method statements answers such questions. All project managers ask questions about time; 'When shall we...?' PERT plays a big role in answering these questions but in the big project they also ask 'how'.

Is 'usually unusual' another oxymoron?

On the other hand, projects within a programme tend to be simple and predictable. As another software-development project is taken on there is no need to have lengthy meetings to discuss how the package is going to be built; the process is well known and a standard project plan already exists. There may still be long meetings but they are to examine the prawn and celery sandwiches and taste the Pinot Noir on the client's expenses, not to plan the project. To build the plan for this new project we need only draw in the standard plan and change the durations to allow for the workload in this particular project. The sequence of steps is unlikely to change. In fact the sequence of steps is often firmly laid out in a published document called a methodology and we'll look at these later.

( Hence the majority of project management is concerned with critical path, method and timing. The majority of programme management is concerned with timing and resource requirements?

Another difference to mention is the shape of the histogram. In the single-project environment the team generally hire in the workforce they need to undertake the project on time. The resources might be employed by a subcontractor but they are nevertheless hired in to do some work on the clear understanding that, when the work is done, they will be expected to move on to another job. Single project workers beaver away in an unmitigated effort designed to put themselves out of work.

A key objective of the planner of a single project is to minimize the number of resources hired in to do the work. If the planner can find a clever way of doing the work and reduce the demand from 16 welders to 14, he deserves a star. If work drifts behind schedule, the first thought will be to hire in some people.

( In the world of programme management the resource levels are much more fixed and static. It takes quite a long time to recruit new members into one of the functional departments and to get them up to speed. It also takes some time and costs some money to get rid of a resource. So the programme-management team's objective is to:

- keep the resources busy 100% of their time;
- keep a long-term eye on the future demand for resources)

Projects managers like to hire as few people as compatible with the required progress. They often hire in a few hands for a few days to help over a busy period. The accuracy of resource planning is not usually sufficiently accurate to plan to the nearest individual resource, it does fine thank you if the team know how much of each skill is likely to be required next month. Generally it is possible to hire in a few extra resources for a short period of time. As long as it's compatible with the rate of progress, project managers try to hire as few people as possible to minimize resource requirements.

Programme managers have a relatively fixed resource pool all of whom have the strange idea that it would be really nice if they could receive a pay cheque each month. The workforce can be expanded and contracted by hiring people in and letting people go (see above, especially if you're in a ravine) but the process tends to take some time. People have to be 'brought up to speed'. It is possible to plan each individual person's time in half-days or even hours. The objective then becomes to keep everyone busy, to maximize utilization.

Therefore project managers tend to want to keep resource numbers down, programme managers want to keep utilization up.

Project managers like low histograms whilst programme managers like smooth ones.

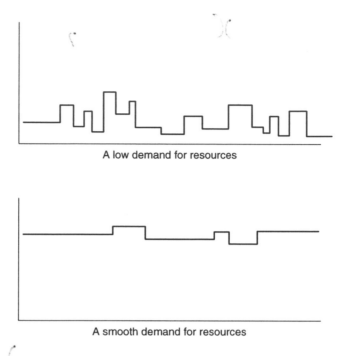

A low demand for resources

A smooth demand for resources

Project managers tend to have simple objectives. I do not say that these objectives are simple to achieve, I do say they are simple to understand. The project manager's job is to build this for that much by then. As long as the oil refinery is finished to specification, to budget and on time everyone should be happy. Everyone will probably be just a little less miserable but that's another issue.

If the price of oil drops through the floor or if a scientist in Lower Serengetti invents a fuel made from earth and water the oil refinery will become useless overnight. The oil company will wring their hands wondering why the sky has chosen this moment to fall on their heads. The project manager will steam on safe in the knowledge that his job is to get it built to cost and budget. Programme managers have to worry about benefits.

Programme managers have to watch the environment closely to make sure that each project's objectives still make sense and still help the organization to achieve its overall strategy. They have to be ready to drop a project altogether, modify some others and introduce some new projects if for some reason the benefits of a project look like being whittled away.

Programme managers keep their eyes on the corporate objectives, which are strange animals subject to interpretation. They will drop a project like a ton of hot chillies if it appears that the project no longer aims towards the corporate goals. This might be caused by the corporate goals changing due to a policy shift, an environmental change or a change within the project.

A policy shift is a polite term for a board member changing his mind for no good reason. An environmental change does not mean it has started raining but does mean that something outside the organization has changed. Plans to build a second ferryboat might easily get dropped if a new bridge were announced by the local government. Some projects drop themselves – if during a pharmaceutical research project it comes to light that the new wonder headache drug has the side effect of creating hallucinations in males with beards the chances are the drug company will drop the project and pass it over to the Colombian drug barons.

Actually this never happens but pharmaceutical firms start many more drug-development projects than they expect to finish. Each project is in a survival-of-the-fittest race during which most will get dropped long before they see the cold light of day.

And finally (we're still talking about that table), there are many tools which handle single projects very nicely but very few aimed at programme management.

Yes, I know that every software supplier claims to deal with programme management but very few do. They all offer 'programme-management functionality', which means the ability to merge files, a few offer the ability to create and maintain a hierarchy of plans which you can navigate through to find the bit you want. Some of the mainframe heavyweight systems offer multi-project management but there is, I am convinced, a range of new tools on the horizon that will deal with the programme sensibly and fully.

> Why is it so much easier to answer questions you ask yourself?

## The difference between MRP and programme management

If projects are about one specific objective and programmes are about many simultaneous objectives, how does MRP fit into the picture? Now I am no expert on the world of manufacturing but I did a little reading to try and understand how these three – project and programme management and MRP – fit together.

MRP is 'material requirements planning' and is to do with a production-flow process. Here we are at the end of the 'project' to 'process' spectrum furthest from the single project. At one end is the one-off mega-project like the Eurotunnel. Such projects tend to be huge, very unusual and rather romantic in a curious way. In the central band of this particular spectrum is a wide range of normal projects which are smaller, more familiar and which fit into neat categories. Right at the end is the continuous-flow process where oil, tomato juice or something equally tasteless flows along a production line  being processed as it goes.

MRP is used to help plan at the process end. To confuse the issue somewhat people in production talk about projects meaning the sort of development work that brings a new product to the market.

When process people talk about projects they are talking about the process involved in bringing a new tomato juice to the market. You dream up a new variety of tomato juice – perhaps it has got Worcester sauce in it already.

> My idea first, Mr Heinz.

You have to take this idea and convert it to a saleable product. You have to set up agreements to purchase the ingredients and get the product passed by all manner of authorities both inside and outside of your organization. You have to design and organize packaging and plan and make your marketing campaign. Maybe you make some TV ads, maybe you employ a freak to shout 'Buy Tomato Juice with Worcester Sauce Now' in between records at local radio stations.

You really ought to be ready with trucks to distribute the stuff and…well, you get the idea, there is loads to do.

Hence there is a project. The project begins with the objective of getting this product out  on to  the market and ends when the production people take over – when the product is being sipped in hotels and bars throughout the land. Manufacturers like to reduce the **time to market** as this is the investment side of the equation.

The quicker the product is got out  onto  the market the lower the investment, the earlier the return and the less chance your opposition have of bringing out a competing product before you.

If we assume for a moment that a manufacturing company is currently bringing a number of new products to market then we have a programme. You could easily regard the research and development activities, the pre-production modification of manufacturing plant and the early marketing work for a number of new products as being a programme. It has all the classic hallmarks of a number of projects calling on the skills of a number of functional departments.

Once the project is over and the product is on the market we can refer to production control tools and ideas like MRP and JIT. JIT stands for **just in time**, which is also known as **short-cycle manufacture**. If you thought that short-cycle manufacture was building bikes for midgets you are beginning to know my mind far too well. The concept here, like most concepts, is simple – car manufacturers are fairly typical. 'Why', they ask, 'should we have a huge stockpile of components like brake drums and windscreens to support our production line?' Having a stockpile is very expensive as you have to tie up large sums of money in the stock, you lose and break quite a lot of it and employees have this annoying habit of accidentally dropping valuable bits down their trouser legs just before limping off home for tea.

You also have to transport and offload each bit before stacking it, moving it about on a forklift truck, losing it, finding it and eventually feeding it down to the production line where the cars are made.

Why do this? The answer is 'Because we want the production line to keep on running. The costs of stopping the line because we have run out of wiper motors is huge.'

This was very true and widely hailed as being very sage until computerization gave us the ability to predict with some accuracy what cars were going to be built and when. All you have to do is to break down each type of car into a list of components and schedule each type of car. The computer can then pump out a list of the components you need and give each a schedule.

Hang on a mo – break down the list of components – how is that done? Good question – you draw a hierarchy like a family tree with a car at the very top and break it down into its major components and then break those down at the next level into small components and so on. Here is a little example:

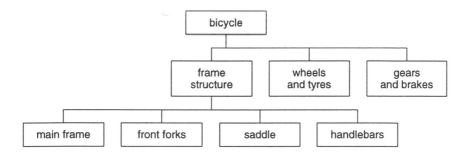

OK, so it's not a car but it is much greener than a car. It is also incomplete but I hope you get the idea. This is sometimes called a **component breakdown structure**. Take this diagram and convert it into a list using the same structure but representing the boxes as headings and you get a **bill of materials**.

```
        bicycle
            frame and structure
                main frame
                front forks
                handlebars
                    bars
                    grips
                    stem
                    locating nut
            saddle
                seat
                seat post
                fixing bolt
        wheels and tyres
        gears and brakes
```

Does this look anything like a work breakdown structure? Well it ought to because the concepts are very similar. There is a difference: A component breakdown structure simply lists all the components that exist within a deliverable. A deliverable is the name for the thing you deliver at the end of the project. The bill of materials lists the same components in a list form.

The work breakdown structure also deals with the work that has to be done to acquire, manufacture and assemble these components. The WBS would have tasks like 'Interview possible saddle suppliers' and 'Place order for saddles' and 'Wait for delivery of first saddles'. WBS diagrams and lists have 'doing' words in them.

This JIT concept and the general computerization of vehicle manufacture has given us a huge range of cars with all sorts of ridiculous names and option packs. When you order your new Fraud Acronym 2.0lt XLS Gti 16V Aerodeck 4WD RWS Econogreen in terracotta

('Another one in brown, Joe') with the two-tone brown trim, air-conditioning and driver's airbag plus the optional ice-warning alert and ski rack, your wise and perceptive selections all go into the computerized ordering and manufacturing system.

Suppliers are advised on forthcoming demands and are expected to deliver the goodies just in time. Hold on there, wasn't that the name of the scheme? No, not 'deliver the goodies', 'just in time'. The suppliers are supposed to back their articulated lorries up to the side doors of the factory laden with the right pieces of car in the right order as required by the production line. If you stop the right lorry on its way to Halewood you could find your ski rack somewhere in the back.

The motor manufacturer has much less of a stockpile, less chance of wastage and theft, less money tied up in stock, less warehousing and is feeling better all round. The suppliers are sweating like mad to keep up. I sometimes wonder if JIT really means that the stockpiles are in warehouses belonging to the suppliers instead of the manufacturers.

---

There is also the JTL technology – just too late.

---

There is another clever scheme called **concurrent engineering** and many people will gasp in wonder that any industry has not always being doing this. If you are not into concurrent engineering, you have a team of designers who sort out a complete design for your new car, bicycle or whatever down to the last nut and bolt before anything solid gets underway. I guess the building industry is one of the worst in this respect. The architect spends hours pouring over a hot drawing board or CAD system, designing your hotel down to the last hinge and bracket.

Once the design is very nearly finished the quantity surveyor takes over and lists all the components in a book called a **bill of quantities**. This lists all the concrete, drainpipes and doorknobs in detail. This book gets sent out to building contractors who submit bids to build the hotel and the lowest price usually gets the job. Work then starts on site in a sea of mud, progresses through the foundations and concrete frame up to the roof. Then the internal work begins and at some point much later the door hinges get fitted.

You might begin to see the slight suggestion that the construction work could have begun long before every hinge was designed. If the design work follows a sensible pattern it will be possible to get some construction work going on in parallel with the design work. This will lead to a faster project overall, giving the client his hotel much earlier. This is referred to as concurrent engineering.

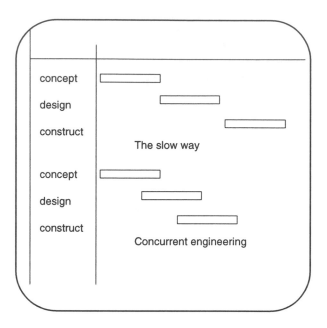

This is also referred to as the 'over the wall' mentality. One group of people is paid to perform some function and to achieve some objectives. The design team is paid to design and their objective is to deliver a complete design package for the new product.

Another group is paid to build and test prototypes. They get the design, moan like mad about those stupid designers who couldn't design their way out of a paper bag, those designers who appear to know nothing about the practical nature of building things and who will be last in line for bread when the revolution comes. After a suitable period of moaning one of two things happens:

1. the design gets sent back as being impractical which starts another session of moaning amongst the design team or
2. the engineers grudgingly get on with the job.

In such cases the objectives are screwed up. The design's team objective should be, like everyone else's, to get a brilliant and profitable product out on the market. The minute you put the designers in one room and get them to design, the engineers in another room and get them to engineer and so on with other groups, you are building walls between the groups.

The Chrysler Viper motor car was brought to market with record-breaking speed because the team were plonked together in a room.

They literally tore down the walls and they all made it work. The driving forces behind getting products to market quickly conflicts with the complex nature of modern development projects and the intricate quality and safety legislation that often applies.

With all this in the background you can see that programme management is going to help manage these development projects through the concurrent-engineering stages, through the setting-up stages, through the prototypes and test marketing, and into JIT production. There is therefore no conflict between project management, programme management and MRP. If you are in the business of making small numbers of expensive things, programme management is likely to be for you. If you are in the business of making large numbers of cheap things, MRP is your boy.

In either case, programme management might be very helpful in the research and development phase as you try to bring new products to market, but it is MRP that will help you to manufacture your new product efficiently.

## No plan is an island

I think that industry has been soldiering on with project-management software for a number of years with a vague, uncertain feeling that the tools aren't really quite what were needed in the first place. I also like to think that Father Christmas exists and that Michelle Pfeiffer would fancy me if she got the chance.

People say that 'these project-management tools help', 'our competitors are into project management in a big way' plus 'the MD says it's a good thing so who are we to argue', but still there is a subtle sense of dissatisfaction. Wordprocessing, spreadsheets and accounting systems seem generally to perform their jobs, doing what is expected of them. Project-management software seems to be left out and feeling a little down. Feeling a little down does not mean stroking a duck.

Part of that dissatisfaction might be to do with the name. Project-management software is a terrible overselling of a disk or three. People actually expect these products to manage projects. What a silly idea – next you'll believe that slimming aids really work and that marketing people and politicians are truthful. There is a very wide range of very competent software systems to help plan projects but none of them could manage their way out of a paper bag.

> How can you tell that a politician is lying? His lips move.

When your innocent and intelligent manager gets given a copy of *SuperTimeProjectWorkbench* Version 6.4d she expects it to help manage the project and it doesn't. She expects it to solve her problems and it does exactly the opposite. It helps find problems; it may help locate problems in good time, whilst there is space to take some action, but it does give you problems. You have to make the problems go away.

Whoever said 'Don't give me problems' was not into project management. That's exactly what a good part of it is all about: finding and spotting problems whilst they are small and in the future so that they can be effectively removed.

So project-management software is a bit of a disappointment just because of its name. Even if they were more accurately called project-planning systems you would still have the vague notion that there is something missing and I think I know what that something is.

In project management the computer is an island at which no bottles arrive carrying messages. It is a lonely little place on which the desert island inhabitant plays his eight records, burns the complete works of Shakespeare to keep warm and models his projects.

The key word in the previous paragraph was 'eight'. No, I am only kidding to see if you're still awake, the key word is 'models'.

Project-management software is about modelling the project. You can model how to put the roof on, how to test the software, when to deliver the printing press, but the project may be totally unaffected by all this modelling. The modelling helps people to see a little way into the future and therefore it helps people make decisions and these decisions affect the project. It is the people who affect the project, not the tools.

Weather forecasting is a modelling process and you would agree that it doesn't affect the weather one jot. When you hear Michael Fish describe tomorrow's weather you decide what to wear to the races or on that picnic. You make your decisions on the basis of the model and so it is with project management.

Assuming you are over eight years of age, you must have seen pro- jects where someone busily plans away in a corner and someone else runs the same project in complete ignorance of the plan. I have even been on projects where the project team get the barcharts out of the bottom drawer when the planner comes to visit. The site team examine politely the carefully drawn and printed reports that arrived in the plan- ner's briefcase and then stick all the plans away when the planner drives off in the company Cavalier.

Compare for a moment the role of purchasing software. Someone decides to buy some trout. The request goes to the purchasing depart- ment, which enters the requirement into the central purchasing sys- tem causing an order to go out in the post and the copy order to appear in the accounts department. If there is no order, there will be no trout. If the accounts department don't get to hear about the order the invoice will not be paid and eventually the supplier will demand their trout back.

> At the Motor Industry Research Association a project manager told me how a water sprinkler system was designed to soak the roads on which they test cars. The water ran back to a pond and the pond quickly became clogged with freshwater shrimps. So they ordered fifty trout to solve the problem, to the pleasure of anglers who work at MIRA, the local trout farm and the trout themselves. The shrimps were less impressed.

The purchasing software has not been an external model of reality, it has transmitted information and played a role. It is in the main line of purchasing. Take it away and it would need to be replaced by a manual system or something else. Accounts software is similar as is any soft- ware that controls a machine.

In projects there is the niggling sense that things would soldier on without the plans, the cash-flow curves, the labour histograms and the splendid, colourful critical-path diagram you produced on the design department's plotter. Somehow the deliverable would be delivered. The worst thing is that you can't even prove that your planning helped. As the project was only done once and for all the proof you have, the plan- ning might have made the project take longer and cost more. You cannot prove that project planning is anything other than superfluous.

These are some of the reasons why project-management software gives people a sense of everything not quite being right. Programme- management software can be different.

In an organization where all work is planned, resources are allocated and assigned to do work, people find out what they are supposed to be doing this week by reference to their computer. Their terminal is still an island but bottles with messages in them are coming and going all the time.

> Isn't terminal a deadly name for a keyboard and screen.

The messages travel across local area networks or back to the mainframe down a wire. The software knows what is going on in other parts of the organization and is kept in touch with plans, changes to projects, assignments and all manner of things which are being added to the system on other islands. Each user finds out what they should be doing by looking at their part of the plan and can report back what they have actually done, thereby affecting other people's plans. The software is no longer an external model of the project, it plays a central role in modelling and in communicating that model across the organization. Now, as your local privatized Electricity Board almost certainly don't say, 'We're cooking with gas.'

Some organizations have achieved this, they are cooking with gas. People do not wander about picking up projects at random and progressing them a little. They put the programme-management tool in centre stage and tell each other what's going on through the system. In one sense it is a system for transferring messages in bottles between the many islands.

Of course this doesn't come cheap. To make such a system work you must plan virtually everything and everyone and this takes time and effort. Given such an investment, which has all the spin-off advantages of a single-project plan, huge additional benefits can arrive. That niggling feeling should go and be replaced by a more appropriate view – 'it's a useful tool'. Plus expectations created by the title 'programme management' are much more likely to be met. What more can you ask?

## Teams

'Teams' is becoming a big word. I get the feeling that England, which is where I happen to live and work most of the time, is way behind the world trend in this respect. Organizations of all kinds seem to be taking their organizations by the scruff of the neck and giving it a severe shaking. The result is sometimes a team approach, which generally seems loads better. You do hear lots of people saying how the team concept, which I'll get around to in a moment, has brought them wonders and benefits galore. You do not often hear that 'we tried the team thing but it didn't work for us'.

I was terrible impressed with the Chrysler Viper. It is one hot motor car, it looks great and is sexy enough to almost make you forget how bad it is for the environment. Behind the car is a revolution in thinking back in Chrysler Corporation. It is a good example of the team approach so let's use it.

In the 1980s Chrysler were not, generally speaking, having a great time as a car manufacturer. They had a typical functional structure with groups of specialists in the many elements of a motor car. They had engine people, suspension people, electrical designers and no doubt, door-lock specialists. They took a very long time to develop each new car, which did enable them to boast, like most manufacturers, how much they had put into developing the latest machine. People with a wary eye would look at the 'new' model and comment that if they had really spent $100 million how come all we can see is a new ashtray and a different front grill?

So when Viper came along they were persuaded to throw caution to the wind and create a project team to come up with an innovative new car. The project manager talked about 'taking down the walls' and he was talking metaphorically as well as physically. They brought the team together to work in one room by borrowing all those specialists for the duration of the topic. All the curious and not well-understood team-dynamics things started to happen and the car came out in record time and is a stunner.

The chairman of Chrysler, Robert J. Eaton, said: 'When people feel empowered and trusted, they are at their innovative best. They aren't afraid to take risks, to stretch goals they set for themselves or to engage in the kind of thinking that leads to real breakthroughs.'

This system of management seems to have rippled down from the car makers to their suppliers and on to subcontractors in other industries. Japan has been at the forefront for many years. The idea of focusing on the client and of bringing together the project team is becoming known as **quality function deployment**.

I was lucky enough to be part of a team that worked on a project like this. It was not a car but a cigarette factory and was therefore equally environmentally harmful. The newness was not so much in the technology of the building but in the relationships between the client, the builder and the contractors. So the team was formed and the team spirit was cemented together as most members lived and worked around the work site for the working week before going home for the weekends.

I have rarely been anywhere more motivating or exciting. We had a great time, achieved a wonderful project in record time and got a telegram from the client congratulating us all. Nearly all the team members went on to climb the slippery management tree more quickly than most. The building is still churning out millions of cigarettes each day.

Of course the teams in this kind of thing cannot contain every specialism and the matrix still exists. The team borrows some key specialists from functional departments on a secondment basis and they work for the team for the life cycle of the project. Other specialists are brought in from time to time to provide specialist skills when needed.

During the life of the project the short-term members come and go with a wistful look at what is a great working atmosphere. The full-time members stay with the project until it is over before returning to the functional department with much more fond memories of the time they spent in a team.

The team approach takes a deal of confidence as the company may sink a great deal of authority and control in the teams and the senior management have to take the sour with the sweet.

Much work that has been done on team theories and psychological profiles, though, has a role to play. The idea is that psychometric testing can help you find the sort of person you need in your team. If you need a book-keeper or someone to maintain records, you need a careful, accurate person. A project manager is more likely to be forceful, creative and a leader. This kind of psychological work tries to keep square pegs out of round holes.

## Definitions of programme management

It is important to recognize that the terms in programme management are loose and have not yet settled down. The CCTA's *Introduction to Programme Management* has already helped. The CCTA is the Government Centre for Information Systems, which is strange. Firstly, whoever heard of a government interested in information? Secondly, how do they get the acronym CCTA out of a title like that? Whatever your political leaning, if you work in a government department, quango or local authority you might expect the CCTA to be an expert source of knowledge on all things computing. If you work in the harsh world of commerce you must apply what they say with a little circumspection.

CCTA originally abbreviated the Central Computer and Telecommunications Agency but the name got permanently shortened in a puff of governmental logic.

The CCTA, especially when it comes to programme management, publish loads of useful stuff aimed at the non-commercial world where it is more important that justice appears be done than actually is done. The CCTA definition of programme management is:

> The co-ordinated management of a portfolio of projects *to achieve a set of business objectives.*

The italics are my own and emphasize that to the CCTA programme management indicates more than multiple projects. The CCTA's *Introduction to Programme Management* talks about defining the long-term objectives of the organization. Once these long-term objectives are established the organization identifies projects that help attain these objectives and thinks carefully about the benefits these projects are designed to bring about.

It advises that the organization set up assorted structures to manage the programme and keep the strategic objectives in mind. The sort of projects are likely to change the organization itself, after all we are talking about relocation projects, rationalization and reorganization projects. So to give the organization a chance to stop and take a look at what has changed, what is to change next and to compare all of that with those highly-significant overall objectives, the CCTA recommend achieving 'islands of stability'. Whilst on an island, the ground is firmer under foot and you are better able to take stock of the past, present and future. Several thousand marks out of ten for the person who dreamed up that phrase.

If you are in a publicly-funded body or indeed any large organization about to go through considerable internal change, the CCTA's publications on programme management are well worth a look.

They do not mention the idea of seeking work through competing for projects or projects that bring new products to market. The most avid readers of the CCTA's publications are more likely to be inviting tenders for work than submitting quotations for work. The CCTA approach to programme management is at its most appropriate when applied to a publicly-funded body. It is just fine when dealing with the privatization of British Rail or the decentralization of the Gas Board.

There are many more meanings of the term 'programme management'. Here are the more common meanings:

## ( *Meaning one: the multi-project organization* )

( Programme management is the directing of a portfolio of projects which benefit from a consolidated approach.

Jobbing engineering companies, software houses contracting for work and many other types of organization run many simultaneous projects, each of which may or not contribute towards the corporate goals. Typically the result of such a project is a deliverable which is eventually delivered to a client for payment. After many delays, the payment arrives and gets paid into the company's bank account thereby increasing cash flow, which is achieving one of the company's objectives. Sometimes the projects are much more directly aimed at corporate goals – opening a new factory or launching a new product spring to mind.)

The common elements of the projects are that they run simultaneously or at least overlap with each other, they share resources and are supposed to generate some income. One project being cancelled does not necessarily change the organization's general direction. Royal Ordnance is fairly typical of this sort of organization. They have a large number of projects in hand in a number of localities. Something between 100 and 200 projects is probably a reasonable average and most of these involve developing some awful machine to annihilate people. Each weapon will be developed into a prototype before extensive testing (without volunteers) and a short production run before delivery to a 'friendly' nation. Why such a friendly nation should want to blow up so many people is a mystery to me.

Such projects might have one or two engineers devoted to the project for a period of time plus a range of specialists whose services have to be begged, borrowed or stolen. Once upon a time Royal Ordnance used to operate on a cost-plus basis, claiming that it was the only way to work on sensitive, quality-related products like guns and missiles. In those days the organization was separated into functional departments like design, prototyping and testing but there was no one especially interested in projects at all. No one, that is, apart from the salesman whose commission depended on delivery and the client who was waiting for the deliverables. The client might be fending off warring factions as he awaits his delivery.

Hence salesmen and clients became projects managers and chased their projects about the organization as the projects drifted aimlessly from department to department gathering costs and therefore income for the manufacturer. Projects accidentally fell out of the door into the client's arms, not so much because they had finished but because Royal Ordnance couldn't find any more believable reasons to do more work and therefore raise costs.

If it wasn't for the fact that Royal Ordnance still make nasty devices, you would say things are loads better these days. They create projects to develop a new device, allocate project managers, build project teams and run a sophisticated programme-management system. These types of programme run for ever and need have no end date. The projects are separate in that there need be no logical links between projects. Whilst they share the same resources, delays in one project need not cause delays in others.

## (Meaning two: the mega-project

The management of a portfolio of projects towards one specific objective.

Programme management can also mean one very large project. The USA's Man on the Moon Project was such a programme. In this sense the term 'programme' indicates one very large project which is made up from a number of components. This term is so American I shall drop the 'me' from programme.)

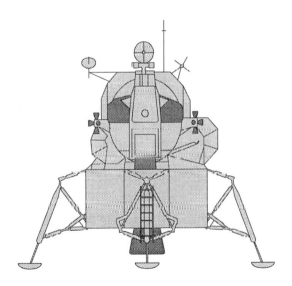

Within the Apollo program there were many projects: the Lunar Lander, the Orbiter, the Launcher and the Control Systems were all projects which were so large, complex and interesting that any red-blooded project-management person would have given their right arm to be involved. Polaris and the Manhattan project (which resulted in the nuclear bomb) are other famous projects large enough to be called programs. Therefore, particularly in USA, the word program refers to a series of projects which make up one large project.

The reconstruction of Beirut is thought of as a program. The war-torn (isn't that a hackneyed phrase?) city is to be rebuilt. There will be many separate projects, each of which will create a government building, a shopping centre, a school or whatever. Together they will become the downtown area of Beirut. You might go there for your summer hols soon. There is a program management team in overall control of a number of project managers, each of which is running a construction project.

The program is usually reflected in the management structure as there will be a program manager to whom the project managers will report. Said program manager, or sometimes program director, will concern himself with recruiting and maintaining his project management teams and on integrating the deliverables of each project into one overall program. In this meaning of program management there is likely to be one physical deliverable.

These sorts of programs end. There will be a time when the overall objective has been achieved and the program and all of its constituent projects are over. There may be a time when everyone has realized how ridiculous the overall objective was and the whole thing has been scrapped, but either way the program comes to an end.

The projects within this type of program are often linked. Delays with one project often cause knock-on effects with others due to logical links between tasks in both projects. For example, if the moon-rocket launchpad project was delayed, it would delay the testing of the moon rocket itself. The Beirut Shopping Mall will be of little use without the water-treatment plant and the new sewer scheme. Such projects may not share the same resources but they are almost certain to be linked through their logic.

## Meaning three: many projects for one client

> The management of a series of projects within an organization and for the same client.

Let's consider a company performing work for many clients with a close relationship with some of those customers. Our supplier might have twenty projects in hand for one particular customer and appoint a programme manager to co-ordinate all the projects in hand for that customer. This programme manager will have a team of project managers each of which is working on a single project for the special client)

An example might be a supplier of components to Ford. Lucas, Girling and Triplex all make a wide range of components that are designed in collaboration with vehicle manufacturers. This collaboration all takes place in Secret as the white-coated ones meet to design the next Ford Escort.

> Secret is a small village just outside Warrington.

Triplex, a part of the Pilkington group, might be working on the next Escort, Mondeo and Transit van all at the same time but the three project teams within Pilkington's will be working with different project teams within Ford. It makes sense therefore for Triplex to tie these projects together into a programme, to give the programme a manager and to let all the individual projects co-ordinate through this programme manager.

Great ideas from one group get carried over to the other groups. Specialists who work part time on all four projects can no longer take time off smoking behind the bicycle sheds with Milly from catering and expect to get away with it by telling all four project managers that they were working on one of the other projects. Clients who don't pay their bills on one project may find the other projects held up pending payment.

Such projects are probably not linked logically but almost certainly share the same resources. They may be carried out by different teams within the contracting organization but probably share the same functional departments.

## (Meaning four: the programme-management organization

The management of a portfolio of projects all of which aim towards the corporate objectives.

or

The co-ordinated support, planning, prioritization and monitoring of projects to meet changing business needs.

Programme-management companies run many simultaneous projects each of which leads towards the organization's strategic objectives. London Underground Limited (LUL) have objectives like 'achieving a 98% record in promptness' and a building society might have objectives like 'having a branch in every high street'. To achieve these objectives entails many projects – property acquisition and refurbishment, staff training, IT support systems and so on.)

LUL have a hierarchical system of developing projects. Each line, Jubilee, Bakerloo and so on, assemble a list of improvement projects that they wish to undertake. This might include new signalling systems, station refurbishment or rolling stock. To get on to a line's list of preferred projects, each group has to dream up some projects, evaluate their likely costs and benefits and compete with other projects. Eventually each line puts forward its own list and competes for the investment money against the other lines. Eventually the cash is allocated and work can begin on the myriad of projects that have successfully passed through this selection process. Ones that fail must wait for another day by which time money might be more plentiful. Sometimes, especially when south of the river Thames, you begin to think that a lot of projects don't make the grade.

In this environment every project plays its part towards the organization's ultimate aims and objectives. Often, as projects are completed, this translates into a revised set of corporate objectives. This is closer to the CCTA's definition.

These projects are likely to be linked both logically and by resources. Projects are likely to provide deliverables which are required by other projects. Perhaps a new computer system for line signalling on the underground will be used in many signalling upgrade projects. Also the projects are likely to call upon the same functional departments and resources and battle through the shortage of these common shared resources.

There you go, four very different meanings for programme management. They share some common factors, for example: they all involve many simultaneous projects, they all concentrate on resources and they all need a multi-project view of scheduling. I therefore propose the term **programme planning** as the planning and monitoring of a number of simultaneous related projects. I think I'll put that in a box so that it stands out a bit.

---

Programme planning is the planning and monitoring of a number of simultaneous related projects.

---

Programme planning is the constant. Whichever of the four definitions above you choose, or whichever additional definitions you might conceive, the likelihood is that once the projects are defined, you will be in the world of programme planning.

## *Types of projects*

There are as many types of projects as there are project managers. Whenever a group of project managers come together you often see a furrowed brow, a raised eyebrow and a puzzled look. Sometimes the assembled project managers seem to be arguing some point intelligently but quite often they seem to be talking completely different languages.

If the group come from the same industry with the same approaches, if the group could swap jobs without much hassle, these problems will not arise. It is when you mix project managers from different backgrounds and different industries that, whilst they use the same words, they mean different things.

Behind this confusion is the type of project they have in mind. The nature, type and approach to your project so conditions you that it can make intelligent conversation hard, if not impossible. As I get the chance to talk to a wide variety of project managers, here are some types of projects. Next time you listen to a presentation at a conference or get involved talking project management down at the pub (have you nothing better to do?) and if you want to make yourself clear, categorize yourself and your projects in these terms.

If, however, you are a management consultant and wish to keep everyone in the dark whilst you talk and get paid a lot...

## *Internal and external projects*

There are two very different attitudes to programme management and these two attitudes stem from the kind of work being undertaken. I class the two extremes as internal or external.

**Internal projects** are designed to change the organization within which they will run their brief lives. This category includes an organization setting up a new payroll and bonus system, a new management information system, a relocation or reorganization project. A great number of government departments are busy dealing with projects that have been brought about as a result of the Citizen's Charter, the Next Step initiative and the drive towards more customer-orientated thinking.

Such projects may use outside contractors but the end of the project is very much aimed at changing the organization. It is this sort of project that the CCTA have in mind and the pages of their two books on the topic make this very clear. These internal projects are driven by needs found within the organization. Such projects have no natural client. There is no external organization that must be satisfied and which will pay the bills for a job well done. Clients have to be artificially created within the company for internal projects and money rarely changes hands.

**External projects** have a deliverable that gets delivered to a customer. The engineering firm making marine gearboxes to order, the printing company bidding for contracts to print things, the software company negotiating deals to write software for their clients – these are all examples of external projects. Such projects contribute to the organization's goals in that they bring home a profit but they do not change the organization undertaking the projects. There are natural clients for external projects and they do pay real money for a job well done.

On the border between these two categories is the new product project. In such a project the research and development team, backed up by the testing group, the prototype department and a market-testing agency staffed by people with handkerchiefs hanging out of their sleeves, dream up an idea for a new product that will set the world alight.

The project team develop the idea, test the prototype product, set up marketing and distribution and then hand the whole thing over to the production department. Is this an external or an internal project? The answer is 'it depends'. It depends on the nature of the product, the relationship between research and development and production and the way in which the project is run. It also shows, as the hairdresser said, that there is no black and white in life, only shades of grey.

## Open and closed projects

This could be called design and designed. Some project managers start a project with a huge bunch of drawings and specifications. These explain in tremendous detail what is to be achieved during the project so that right from the outset everyone knows what is to be done. OK, there might be variations and change orders but these alter the previously rigid definition of the deliverable. Such projects are 'closed'.

Other project teams start out with a very vague brief indeed. I am involved in one such project now – we are creating a programme-management software system. The design has been and will continue to evolve right through the process. There will be a very wide range of dates when we could decide to launch the first version and start on version two. We can and do adjust the balance between the scope of the work and the timescale.

The difference is between designing a hotel and building one. The design team have to design a hotel, which you must agree is a very open brief. They will be told how many bedrooms it should have, the conference facilities and the leisure club requirements but how long will the design take?

No one knows. Set a target and you can bet the design will take 10% longer than target because people are exactly that good at balancing scope and quality with time. More time might result in a better hotel. Once the design exists the deliverable is defined and the project becomes 'closed' – a timescale can be set because the scope and quality are defined. Give the builder more time and you will get exactly the same hotel a little later. Another open project was the first Everest ascent.

## Physical and non-physical

I think I'm in love with these two words: physical deliverable. It is a lyrical, singalong type of phrase; it is an incantation, a mantra evocative of mass emotion pent up and then, orgasmically, released.

> This is just my attempt to get into Pseuds' Corner in *Private Eye*.

Physical versus non-physical is an obvious difference but has subtle implications. Builders, civil engineers and mechanical engineers all deliver physical deliverables. Great lumps of tin and concrete called buildings, bridges, tunnels and trains are all the very tangible physical deliverables of these kind of projects.

People find it so easy to get motivated towards such a target – you can visualize it and many project teams understand the benefits of having a model of the thing-to-be sitting on a plinth in the foyer. This, the model silently says, is what we are here to create. Apart from the pay cheque, this is why we come to work. One day the team members will stand back and admire the thing-that-now-is and think proudly 'I built that'.

No such luck for the non-physical project team. They have no mental picture of the thing as there is no physical thing to imagine. Software people suffer from this as do some research-and-development teams. The thing-to-be will be contained on a disk or three, each visually indistinguishable from any other disk. The end objective might be a report containing thousands of neatly desktop-published words. Unexciting is a word that underestimates the lack of impact.

Monitoring causes a wee problem. Whilst builders count bricks, tunnellers count feet and engineers count welds, what does our poor old software engineer count? People involved in non-physical projects dream up ingenious measures which they can use to measure progress in some artificial way. Lines of code has been used and metrics is an approach which seems to be gaining credence in software. But compared with physical projects, software is like building a four-bedroom detached house inside a large cardboard box. You cannot see through or over the box but you can shout through the wall 'How's it going?' and get responses like 'Fine'. On a day bearing little connection with the planned end date the box will be removed and there will be a petrol filling station just as you didn't want it to be.

What tends to happen in software is that the team set up all sorts of phases and stages under the guise of a methodology to surround the actual code-writing bit. These, to an engineer, are temporary works – things that you create to enable you to do what you really wanted to do in the first place. A software prototype is very much like a scaffold in these terms.

In many of these areas, planning software fails almost completely. You can plan a closed project with a physical deliverable because you know what is to be done and can make some pretty good estimates of how long it will take. The idea of breaking the work down into chunks works quite well. Because the scope of the project is fixed, time becomes the key issue.

But given an open project with a non-physical deliverable you may as well throw your planning software in the bin. It will only distract you from the true goals. You'd be better off motivating the team and there is little more demotivating for a project team with an open project than a fixed project plan. Such a plan sets time above all other objectives, effectively saying 'we'll do what we can by this date' which may be exactly what you do not want to do.

## Runners, repeaters and strangers

These terms do not refer to athletes, rifles and people that you should not accept sweets from. They do refer to types of projects and I find this classification useful. This classification is based on ideas penned by Dr Ralph Levene, Head of Project Management at Cranfield University. He has specialized in the field of project management within large organizations since before the term programme management was dreamed up. The idea is that you can take each project and classify it into one of these three classifications. The resultant classification might well affect the way in which you run the project.

### Runners

These are projects which happen all the time. There are nearly always a few of these types of projects ongoing – they might be your bread and butter. Your organization is probably well set up to deal with such projects and they rarely present major challenges. They are low-risk projects.

*Repeaters*

These projects happen a little less frequently and are a little out of the ordinary. They are similar to the mainstream 'runner' projects but have enough variations to make them worthy of a little more attention. Being a little odd the organization may not be so well set up to deal with them and therefore different groups might have to contribute. The risk with these slightly less certain projects is more significant.

*Strangers*

These are the one-offs of the project workload. The things which your organization has little or no experience of. The organization is certainly not geared for this kind of project and therefore it might involve many different interests and functions. There is likely to be a high risk.

Here is an example. Imagine you are a company producing and publishing popular music. You get groups of not especially spotty youths together, cover their faces with makeup, get them to leap about in a weird location whilst some other people sing and play instruments similar to the ones being carried by the youths. You record all this on video and audio tape and release the whole lot on CD, cassette and DAT. You aim for the top ten, *Top of the Pops* and the record-buying public. This is your business.

A runner would be yet another pop record. It follows the normal pattern and is fairly predictable. The location might change, the youths might be male or female, spotty or not and the music might be brilliant or unrecognizable. You've got people ready and willing and experienced in running projects of this kind. You know the sorts of things that are likely to go wrong so the risks are known and small.

You decide to record a live opera and release that. This means setting up recording equipment at the opera for both video and sound and then editing the tracks together into a more adult package. Marketing will be different, editing and recording will be different but the technology is much the same. A repeater like this will need a project manager who can adapt, adopt and improve. The risks are greater simply because they are unknown. You're going to have to use some new people to get the recording set up in the theatre and working on time.

Then you decide to run a concert at Wembley. This is something completely outside your area of knowledge. It is a stranger and you will need new talents and experience if you are to bring this in on time and to budget. The risks of abysmal failure are unknown and therefore high.

You can see that the team, the contingency and the risks of each type of project vary significantly. You would be foolish to treat all three types of project identically and make no provision for the increased dangers and problems of less familiar projects.

Bringing this sort of thinking to the attention of your superiors can do wonders for your office credibility.

You might even recognize in your methodology that projects can and should be categorized and that the procedures you use should reflect their status in these terms of familiarity.

## *What is the net benefit of programme management and who gets it?*

Putting together and operating a programme-management environment is an expensive project in itself. So what do organizations get for all this effort? The next few paragraphs have two objectives. One is to explain what you can realistically expect to get from a programme-management system and two is to provide a bag full of good points to help you justify your proposal if you decide to travel the programme-management road.

Let's first take the organization planning its work and allocating resources in a consistent way. There might be many planners but they all:

- use the same planning software on compatible hardware;
- use a consistent task-description system;
- use consistent names for resources;
- use the same calendars;

Most software systems allow some mechanism for merging together a number of discrete plans. The idea is that you take a few plans on a few computers, somehow get them onto one disk and then merge them into one large plan. I'll write about getting the plans together on page 147.

> 'Discrete' means 'separate' and does not suggest that the plans will look the other way whilst you muck up your plans or adjust your underwear.

You can then display and print overall histograms showing demand for a specific resource or type of resource across the whole workload. Useful.

You might select to see all the design work in all projects although this assumes consistency in planning task descriptions, resource names and so on across all of the projects. You could ask to see a barchart of all the activities with 'design' in the description or for all activities that use a resource called 'Tarquin'. This would show you the workload made by many projects for that department, resource or individual. This should be all in one simple, useful report with which you can enhance your reputation. It would be as useful as a constipation cure after a prawn vindaloo if you don't have a resource called Tarquin.

Many people plan their projects most of which have tasks involving design, prototyping, testing and so on. Once all the work is planned and the plans consolidated, like tasks can be grouped together and passed to the many interested parties. The design tasks go to the Chief Designer, the testing tasks go to the people in charge of the testing facility. It sounds simple and it works well but it does take some energy and commitment.

OK, the first return on an investment in programme management is the ability to predict resource requirements across a lot of projects and to draw together the many demands for the individual skills or resources.

A view of what is required and when makes for better planning. There are three benefits of being better planned and these are discussed below. There are downsides as well.

## (Advantage one: people think ahead more

When there is an atmosphere of planning hanging in the air people do seem to think more about what they intend to do.)This is partly because if they don't get the barchart on the boss's desk by Friday they are for the high jump and partly because they actually get a smoother and quieter life. Planning doesn't get things done, nor does it avoid all problems, but it does help bring up problems when there is still time to avoid them. Compare for a moment the following two scenes. Spot the balls up and win a prize.

## SCENE A

### Location

An office full of piles of old files; stacks of papers on the desks; phone lost beneath some drawings; bin full of papers, old crisp packets and a suspicious-looking apple core; overflowing ashtrays. A sign on the wall says 'You don't have to be mad to work here but it helps.'

### Characters

*George:* Design Department Manager (DM). He is gravitationally challenged; needs a haircut; his shirt bulges over his expansive stomach revealing a not-very-clean vest; he smokes incessantly.

*Mary:* Project Manager (PM). She is much more presentable, neat, tidy and bespectacled. She carries a neat folder containing a few key documents and she looks as if she had time this morning to prepare for the day and for this meeting.

*PM (enters George's disorganized office, glances around and hides her disapproval):* Hi George. I've got the specification for the Mark 4 approved at last and it's ready to go into the design phase.

*DM (starts a fruitless search under bits of paper and through piles of files):* Oh, right, Mary. The Mark 4. Yes. I've got the draft specification round here somewhere. I can't understand why the board take so long to think about these things – what do they know? I hope you're not in a hurry for it.

*PM:* Well yes, really we are. If we don't get a design proposal in front of the client before November Fifth, there will be fireworks.

*DM (starts to laugh):* Fireworks. November the Fifth. Great. Ha ha. Sorry, no chance. I've got four of the guys working on the revisions to the Winchester job, one on a training course and one off sick. The team I had working on the nose cone have just been taken off it to do some secret rush job which even I don't know about and my secretary has left. I'm not sure we could start it before the fifth. Why didn't someone tell me this job was coming in? I could have hired in some contractors.

*PM:* It is a priority job, it's worth over four million pounds and there could be loads more work from this client. Can't you do something?

*DM:* More work. Do you think I want more work? I'm snowed under as it is!

*(Dissolve.)*

**SCENE B**

**Location**

A neat, tidy office: barcharts and histograms are displayed on neat pin-boards; filing cabinet shows project names; the phone has its memories pre-programmed with other department heads' numbers; there is some orange peel in the bin; the PC is running showing some project plans; a sign on the wall says 'THINK AHEAD'.

**Characters**

*George:* Design Department Manager (DM). He is slim, healthy and relaxed looking. He looks like he knows where he is and where he is going. You would hate him if he didn't seem such a nice chap or if he had a digital diary.

*Mary:* Project Manager (PM). She is equally presentable, neat, tidy and bespectacled. She carries a neat folder containing a few key documents and she looks as if she had time this morning to prepare for the day and for this meeting.

*PM:* Hi George.

*DM:* Hang on a sec, Mary. Let me just save this file... Right, how's it going?

*PM:* I've got the specification for the Mark 4 in for approval at last. I know that it is a week behind schedule and I don't think we can expect to catch the time up. It should be ready to go into design on the 3rd October. Will that be a problem?

*DM:* Well, let's see. Let's pull up the workload plan for October and November. Umm. We'll be overloaded in the second half of the month but I should be able to pull in a couple of freelancers to help out given this much notice. Yes, it'll be fine. Thanks for letting me know.

*PM:* That's a relief – I hate to bring bad news.

*DM:* I can deal with bad news – I can't deal with late bad news.

Spot the differences? Yes, you get the prize for spotting that the apple core had become some orange peel in scene B. Things still go wrong but the people work together to solve them. They have time to plan because they take time to plan.

## (Advantage two: people communicate better

Once you have a plan you have a means of telling everyone who is vaguely interested what it is you all plan to do. A programme plan is different to a project plan in that it covers many people's work and is the result of many people's effort. The plan is a centralized source of information letting everyone know what everyone else is planning to do.)

I take the view that a plan for a single project should be the result of a thinking ahead process involving the project team. I do not like to see project planners working away in glorious isolation, coming up with a plan and then issuing it to the people that are really going to make it happen. Plans are a statement of intent and must be owned by the people who own the project. Even on a single project I believe the planner's role is to interpret the ideas held in the heads of the project team and put them down on paper. It is not the planner's role to decide and tell people what they should be doing.

This is doubly the case in a programme environment. If a planner or two is sent off into a planning office to come up with a plan for the whole works, it will probably not work at all. Projects people will reject the plan as it doesn't fit in with their thoughts and because they don't 'own' the plan. Departmental managers will quickly show the plan to be full of holes and prove it to be unworkable. The worst crime would be for the planners to decide which resources should be allocated to which tasks – a move very likely to antagonize a departmental manager who has the strange idea that he is in the best position to understand who is best suited to which task.

The planner's role in a programme is to help the project managers plan their tasks, help the departmental manager plan their workload and to spot conflicts in good time. The plan should also tie the many project plans and resource plans into one programme plan. This programme plan might be fairly large, it might be huge, so the planner should also help people set up their computer systems or paper systems so that they get the information that is relevant to them.

So, given a plan that everyone has contributed to, everyone is involved in, everyone 'owns', it is possible that people will bring their part of the programme plan up on screen regularly to see what is going on, who is doing what, what problems are developing and what is in their in-tray. They might even think that the planner is a good egg.

## *Advantage three: you have something to monitor against*

Things never go to plan. Things go their own sweet way in complete ignorance of your plans. The weather does its thing without reference to the BBC or the met office. If the wind feels like blowing from the east and bringing huge black clouds over your picnic, no amount of arguing that the weather forecast promised bright sunshine will send the clouds away.

Perhaps an even better analogy is that a journey will take the time it takes whatever your schedule says. No amount of wittering on to the airline staff about your itinerary and how it says you will arrive in Bogota at 15:40 will mend the plane's engine and get you going. You are wasting your time so you may as well enjoy it.

What the itinerary does tell you is that you are late. Not late as in 'the late Ernest Hemingway' but late as in running behind schedule. You know that there has been a delay and that this means you can expect to arrive in Bogota late. Now you know this is likely to happen you can call ahead and tell the people expecting to meet you. Without an itinerary you don't have any means of knowing if you are on schedule, miles ahead or hours behind.

In much the same sense a programme plan is an itinerary that allows people to keep an eye on what is happening and to see if those things are happening on schedule. The plan becomes a useful yardstick against which you can monitor progress and report back. On the positive side you can predict problems, change the future plans to accommodate the revised schedule and avoid too much chaos.

On the negative side you are giving ammunition to the senior management to come and batter you around the ears for being late. It is a regrettable side effect of planning that because you take the trouble to plan your work and publish your plans to let everyone know what you are doing, you lay yourself open to criticism. In the worst organizations people get told off for being behind schedule and therefore avoid issuing plans of any kind. The penalty for not producing a plan is smaller than the penalty for being behind schedule. Leave this next paragraph open on your boss' desk:

> In project management it is not necessarily a crime to be late, but it is always a crime not to know you are late.

What a shame this is too long to be a bumper sticker. You should be respected for planning, encouraged to try hard to make the project move forward rapidly and effectively and supported when difficulties emerge.

OK, this is the real world and many people plan the impossible, schedule the unlikely, laze about whilst the project goes wrong and should be first in line when the flesh-eating monsters invade from the planet Thrugg. Most poor managers don't plan. Most good managers plan well. Say thank you to Geoff Groom of The Projects Group, part time project-management guru, for this other little tortuous gem:

> If you fail to plan, you plan to fail.

Now that could be a bumper sticker.

There is another argument for having a plan – to prove things are OK. Take this little story:

In the absence of a plan a pinstriped person with half-moon spectacles and a smattering of white hair will get the idea into his head that you are miles behind schedule with your project. This idea becomes really significant when people at the MD's golf club (for it is he with the suit, specs and shiny hair) poke fun at the speed at which his new factory is going up down the road.

Regrettably this bit of fairly friendly fun-poking coincides with his eldest son being caught with a small amount of a proscribed substance at a rave, his daughter announcing that she is moving out and his dog catching Distemper. Distemper is the name of next door's rabbit.

This combination of events does little for his temper, his son's temper and Distemper and he takes it all out on you. He gives you a carpeting in front of your mates when you bump into him in the foyer one day. In between assorted rude words he puts forward the view that the factory is going to be weeks late and anyone with an ounce of intelligence can see that is the case. Mentioning metrication and suggesting that he means 2.2 grams of intelligence is possibly the worst thing you can do. Producing a reasonably detailed project plan marked up with progress and showing a healthy state of affairs is amongst the best things you can do, as long as you don't do it in front of your colleagues.

When you are doing well, and I know you expect these times to be few and far between, a plan can help a lot.

## Disadvantage one: programme management can easily be abused

There are some potential uses of multi-project planning which everyone can quickly and easily gain just for the price of a little co-operation and helpfulness. You can share and enjoy. When I say share and enjoy I mean that all the project managers and project planners do loads of hard work so that the senior management can have these new reports on which they can base their decisions before taking the rest of the day off. You do your share so that they can enjoy the golf course.

Sometimes a programme-management environment becomes one of the worst forms of management around because it has become, or even was designed to become, only a reporting mechanism.

In such a case, is there a benefit to the poor old project-management person somewhere lurking? Yes there are loads:

- The project managers learn to use the software.
- ...Err...
- That's all.

(Don't run off with the idea that that is all there is to programme management – there are loads more benefits. It's just that there are many simple direct and visible benefits, to senior managers, not project managers.

The danger is that senior management will worry about project management, read about programme management and decide that programme management will help them along. Next they create a working party, committee or even a new department to examine the topic. They go out and select some tools, set up a methodology and install it. To the average project manager this is just extra work from which the project manager does not benefit.)

*A clip on the ear*

Frequently the only realistic answer to a normal project manager who asks 'What do I get out of this?' is 'A clip round the ear'. This is because most senior managers adopt the police approach to project management which is:

Anything you say may be taken down and used against you.

The senior management adopt an approach to project management which is really a fault-allocation system. They get you to plan what it is you and your hard-working team are undertaking to do. They press you to do it more quickly and with less money and then expand the amount of work involved. Under the pretence of wanting enterprise-wide resource reports you have to plan your work down to the nearest half person-hour using named resources and a structured task-description system. Every second Tuesday you deliver a copy of your up-to-date plan to someone or other who merges your plan with many others. The management then list the projects with the jobs deepest in the excrement first – and then set about battering everyone whose job is on the list.

Project managers complain that 'project management is just the latest fad from the fifth floor [them, upstairs, head office] so I'll have to go along with it. It means more work for me and what do I get out of it?' There is that reasonable question again. The answer is probably 'continued employment'. The planners do, at most, the minimum possible to satisfy the demands of the management. Management, armed with this new data, list the projects in order of greatest slippage or greatest over-budget and give those projects' managers a hard time.

Such a system is not 'owned' by the people; it is not respected; it is not seen as an aid, more as a pain. Such a system has missed the point. It is a criminal justice system set up by burglars, policed by safe breakers and judged by murderers but one where the hangman wears a pin stripe suit and has a rosewood effect desk with leatherette trimmings.

# Managing the programme

## Programme definition

In *Project Management Demystified* I rattled on for some time about the importance of setting objectives as clear as possible for each project before it gets underway. I'll summarize the point here because it so vital. But before doing so, and because we are here involved with programme management, let's start at the top with the long-range goals of the organization if they are relevant. If your organization is the sort of company that sees programme management as changing the organization itself, then your overall objectives will set the general direction into which all projects should neatly tuck.

If your organization is in the business of performing projects for clients and the only common factor is the projects' contribution to profit, this concept is irrelevant. Therefore, where the programme is concerned with changing the organization, the objectives of the company can be translated into projects which are aimed towards achieving the long-term strategic goals.

The projects become objectives for the project managers that are going to run them. You cannot expect a project manager to keep an eye on the overall fit of the project into the corporate goals as the project manager is far too busy and too fired up to do anything other than push on with the project. When you are running hard and doing well in the 100-metre sprint, you don't want people to start asking if you should have entered the synchronized swimming instead.

The senior management should adopt the role of checking that every project is moving towards the overall objectives of the company. As the external environment changes, the needs of the corporation change and some projects become superfluous or even detrimental to the company.

An example: the company has a stated policy of opening a branch of its building society business in every main town. This leads to a number of projects: one is to locate suitable premises in each town, another is to develop a corporate image in terms of interior decoration and staff clothing, another is to have the fixtures and fittings built to the house style and, to put all this together, there is a building and shopfitting project in every town. New shops need to be acquired in a number of towns and one project involves those most maligned of all living creatures: estate agents. They are employed to find excuses to explain why they haven't yet found and purchased a suitable property.

The various projects are well underway when the announcement of secret merger negotiations with another building society come through. Suddenly the 85 shop units that belonged to the newly merged company all need modifying to the new house style as well but 25 newly purchased premises are no longer required. Some old projects get scrapped, some new ones are formed and some go on in much the same way as before. An external change affected the justification for many of the projects.

Most sensible organizations have a **project definition form** of some kind with which to start off projects. This is a formal submission to the programme board to proceed with a project. The project definition form identifies the objectives and purpose of the project, the time and cash budgets, the key personnel and the risks involved.

The programme board may approve the form, thereby giving the project manager the authority to spend the money from the budget. In many companies it is very hard to spend money without a project reference number and these reference numbers are handed out when the project is approved.

After signing off a project definition form the board and the project manager have a reasonably similar understanding of what the project aims to achieve. In the absence of such a statement everyone has different ideas of what to expect and therefore most end up disappointed, however brilliant a job the project manager made of the project.

It is typically hard to carry out a fair assessment of a project's costs and even more difficult to measure the likely benefit back to the organization itself. Part of the problem is that right at the outset of the project you have so little knowledge about the project that you may as well read tealeaves for an estimated end date for the project.

Often approval is given in stages. For example a project may be approved through to the end of the design phase or through prototyping. At the end of that stage, the remainder of the project is submitted for approval to proceed.

**Justification** and **viability** are terms used to measure or compare the value of the project. There are some projects which, no matter how wonderfully they are carried out, will result in a loss. There are some cashcow projects which, no matter how stupid the project manager pretends to be, will result in a significant improvement in the company's affairs. There are a couple of simple tools that help evaluate a project's viability and I'll throw them in with the price of this book.

The first idea compares a number of alternative projects by examining three factors. The idea is that there are too many projects in people's minds for the organization to run, so it must select a few with which to proceed. This is a sort of graphical representation of the 'survival-of-the-fittest' approach.

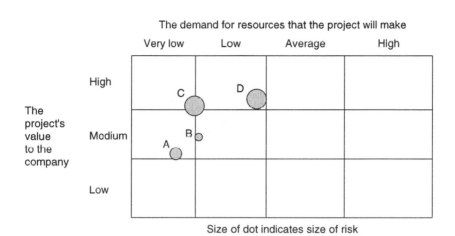

Size of dot indicates size of risk

The three factors being compared are:

- the value of the project to the organization;
- the resources or cash it will absorb;
- the risk of the project not reaching its objectives with those resources.

The small circles in the top left-hand corner are the projects most likely to proceed. Projects that result in large circles in the lower right-hand corner are doomed to the great critical path in the sky.

None of this is very scientific. Whilst you might be able to quantify the cost to the company fairly accurately, you will have to use considerable subjective judgement to estimate a single number for each project indicating its risk and its value to the company. As with most approaches like this it is a comparative process and therefore as long as you use the same values and judgements for all projects, the comparison should be fairly fair.

It is also a neat way of showing the senior greybeards why you are recommending whatever you are recommending. What you are really recommending is an increase in your pay cheque.

There is another approach which is used where the result of each project is to increase the companies market share. I have in mind those organizations that develop new products, enhance existing products, run advertising campaigns and do countless other things to convince you that you really ought to buying their petrol, washing powder or computer. This is mostly completely false as we all know that all petrol, washing powders and computers are actually identical apart from the wrapping.

---

## A little side issue about washing powders, pixies and gnomes:

---

Once upon a time there were two brands of washing powder, one made by pixies and the other by gnomes. They both had roughly the same market share but they both worked hard all day to gain a little extra market share from the other.

One day, a fairy visited Rapunzel, the marketing director gnome who lived on the top floor of a huge mushroom, and said: 'Hey dude, have I got a wizzard idea' for that's how fairies in this story speak. 'We should launch a new washing powder with a different brand name and print 'gnome' in tiny letters inside the bit of the box you tear off. Then there will be three brands on the market and they will all have around 33% market share each. The trick is that we will own two of those brands and therefore actually we'll have 66% of the market!'

'Brilliant' said Rapunzel who had been watching *Star Trek* on TV, 'Make it so.'

The fairy's idea worked well for about a week until the pixies realized what had happened and brought things back to 50/50 by launching a second brand of their own. The gnomes and the pixies went on inventing new washing powders one at a time, each time taking and giving back a little market share. Both were too scared to stop. Both pretended that each new brand was new, exciting and represented a scientific breakthrough. The claims to be better or different became ludicrous as the gnomes and pixies had to dream up something special to say about their new product.

Moral: That is why we now have forty-two brands of washing powders from the same two manufacturers.

Every project which aims to increase market share should have a beneficial impact on the organization. Most projects have the

objective of increasing profits but there may be other objectives like the following:

- increasing the customer list;
- increasing customer satisfaction;
- making the company more efficient;
- improving the product's manufacturability;
- improving the brand's image;
- improving morale within the company;
- decreasing wastage;
- decreasing shipping costs;
- increased knowledge through information systems.

Profitability depends on a whole host of other factors – it is not simple to calculate. You might have a brilliant wheeze and set up a cheap and effective project that should increase sales levels by 10% but coincide with some consumer scare about your type of product. Nevertheless you can compare the cost of the project with the effect you estimate the successful outcome of the project will have on market share.

This can be very revealing as, armed with such a comparison, organizations can sensibly evaluate the rights and wrongs of each project or of a combination of projects. The question becomes: Should we spend £x million on a new product on the basis of a 3% increase in market share?

The graph used to evaluate such questions looks like this:

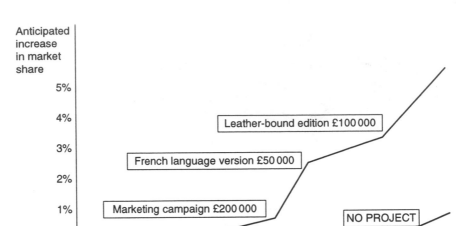

Here there is a suggestion to launch a £200 000 marketing campaign which it is estimated will result in a 1% increase in market share in the first quarter of 1996. Later, a French-language version of the product is planned and a deluxe leather-bound edition is planned for the second quarter of 1996. Each one is expected to have an effect on sales levels and these are shown on the graph. The light grey line shows what the planner expects to happen if no projects are undertaken at all. You can see that you can demonstrate quickly the effect of one or more projects. What is much harder to do is calculate the expected rise in sales.

There is a personal issue here. When you have an idea for a project that will increase sales, you naturally get 'behind' the idea and become a salesman promoting this new idea of yours. It is a human thing to do. The trait in humans that bubbles to the surface in such situations is called optimism. You will inevitably tend to look for the good things, ignore or underestimate the dangers and oversell your idea.

I'm not getting at you, it is just a normal human thing to do. By exaggerating the benefits of the project you set yourself a target to aim for. If you sell an idea for a new product to your board of directors on the basis that it will increase sales by 15% and then later achieve 12% they will be unhappy. If you sell an idea for a new product to your board of directors on the basis that it will increase sales by 8% and then later achieve 12% they will be very happy little chappies.

So the trick is to find the lowest targets that the men in grey suits will accept before giving you the go-ahead. Too low a target and the project does not seem worthwhile. Too high a target and you will never achieve it and will disappoint. Watch your back.

# Structured diagrams: WBS, PBS and OBS

## Introduction

There are three structures in common use within the area of programme management and I want you to get them clear in your mind. There does seem to be confusion or at least different opinions of what these things are all about. They all have their value, they all have their dangers and they all have their costs.

We can refer to them all as structured diagrams. I'll talk a little about structured diagrams in general and then take a look at the big three – **the product breakdown structure (PBS), work breakdown structure (WBS) and organization breakdown structure (OBS)**. They are all made up of little boxes joined by thin lines which, you must agree, doesn't help to separate them from each other or the national telephone network.

They are all attempts to get a better image of a complex thing – like a machine, a project or an organization. Producing and showing everyone a structured diagram has a number of benefits:

- You get a clearer idea about your target. Getting a group of people to spend some time producing a breakdown structure has huge benefits. The very act of trying to figure out how the project is to be run, what needs to be done and how the components fit together gives those involved a much clearer idea of the task ahead.
- You have a degree of agreement. Structures give a shape to the project, the organization or the product and, because they are the result of a useful and searching discussion, represent or define something. Vague edges are cleared up by the preparation of structure. Questions like 'Are we supposed to be doing the helicopter landing pad?' and 'Does the pilot's training school come under training or support?' are answered in these diagrams.
- You have a means of communication. Whilst the greatest benefit goes to those who prepare the diagram, the diagram is an excellent way of telling other people what is going on, what the project is made up of or how the departments relate to each other.
- You have the beginnings of a team. Because you have got the key people together to work on this document they begin feel part of a team set up to do something. Whilst there is a great deal more to team building, the joint preparation of a diagram is a great start.
- It is a structured way of thinking ahead and thinking ahead is about the most fruitful way project managers can spend their time, apart from eating apples and bananas.

For this last reason it makes loads of sense for the project team to prepare the breakdowns themselves. They will enjoy it and gain enormously from the experience. If you get someone else to produce the diagram so that you can show that the diagram has been produced, you are wasting a great opportunity. Good managers don't (or should not) ask for a WBS because they want to see it; they should ask because they want to know that you have prepared one.

When you have assembled the team and they are busy working away with the diagram, let the discussion run fairly free. When I say fairly free I don't mean onto the price of cheese, but there is no reason to stop the team thinking and talking about how some component will fit into the picture, how it will be transported, who is going to make it and a thousand other questions that begin with 'How'.

This sort of discussion is useful and beneficial – the team are finding and solving problems whilst there is still loads of time. Every little issue that is raised and dealt with might have become a huge issue when there is little or no time left.

To hear people talking about the 'how' of the project might be surprising because there is very little 'how' in any of the structures. They tell you what goes into the final product, what is to be done to get there and who is going to do it. They don't say much about how things are going to be done.

For that matter neither does PERT. It is only in people's minds, notes and documents that you read how that spire is to be fitted onto the church, how that component is to be transported or how the software will be tested.

## Drawing structured diagrams

Some time ago a very clever man in 3M invented post-it stickers. The story goes that he accidentally invented a glue that didn't dry out and everyone said that it was a very silly idea. He made up some little yellow stickers and gave them to the 3M secretaries. They spread like a sumo wrestler's waist line. Despite, according to the story, the resistance of the senior management this became a product essential to modern business. For a start Taiwanese monitor manufacturers would have to rethink their designs if the border around the screen wasn't destined to be covered with little yellow stickers.

> Why are they nearly always yellow? Answers, not on post-its, to the author please.

Another very clever man, who is unknown and a bit of an unsung hero to me, used post-its to create a structured diagram. They are just great. You can scribble down words, jobs, tasks, phases and whatever on those little bits of paper and stick them all over a handy blank wall. You could stick them all over that sumo wrestler in the last paragraph but he might get upset and you should never upset anyone twice your weight.

> If a sumo wrestler isn't twice your weight, start exercising now.

The obvious advantage of post-its over a pen and paper is that you can chop and change willy nilly, always assuming that willy nilly is happy to be changed. You can move bits about, merge bits together, try out different arrangements, add in new notes until you run out of post-its, wall space or time. White boards are pretty good as well.

The advantage of using a handy vertical surface is that the people working together on the diagram are standing up and all looking at the diagram the same way up. If you try doing it on a table the chances are that some will have an upside-down view and others will be wandering around peeking over other's shoulders and seeing the growing diagram from the side. Verticality keeps things in line and helps keep people awake and interested.

The advantage over bits of card or paper is that when Mrs Tomlinson comes in to vacuum out the offices later that night, or when the chief engineer pops in to see what's going on, or when someone opens the window for a breath of fresh air the diagram does not self-destruct in a pile of little bits of paper.

The advantage of post-its and a wall over a computer-based system is to do with the team issue. There are computer-based systems and these are very good at representing and printing out diagrams, but I would always start with the group, the stickers and the wall. Once it is sorted out send someone off to fight with the software until the diagram within the computer looks like the one on the wall. If you start with the computer you will do it alone, or quickly lose the group approach, as everyone tries in vain to crowd around the screen and whilst you try to remember if it was Control-Alt-backslash to add a new box. If you've got one of those display gadgets and can throw the screen image up onto a wall, that might work. Don't forget to stress that there is as much of the value is in the doing as there is in having it done.

These three structures fit together. If you have a diagram which shows the work to be done (WBS) and a diagram showing who is available to do the work (OBS) you can map the two together to show how responsibility for doing work relates to the work that is to be done. This mapping of organization onto work is called a **responsibility assignment matrix** and I'll deal with this last so that you have a chance to understand the other diagrams first.

Let's look at the three different forms of structure.

## *Product breakdown structures (PBS)*

Let's start off with the product breakdown structure (PBS) which some call a component breakdown structure (CBS) because they find it to be a more descriptive title. For these pages I'll stick with the more popular, if slightly misleading, PBS.

The PBS is designed to show the components that make up the deliverable of the project. This does not show who is doing the work or even what work needs to be done, it shows what the objective of the project is made up from.

PBS diagrams come from the engineering industries where they need to get a list in some sensible structured form showing all the bits and pieces that go into a product. I saw one at a Formula One racing-team headquarters and it listed every nut and washer, bit of pipe and fitting that goes into the modern Grand Prix contender.

Under headings like 'engine', 'suspension', 'gearbox' and 'ignition' came lower-level headings like 'pistons', 'hubs' and 'spark plugs' and under those were even smaller components.

There is that key word again – components; the PBS shows all the components that go into the deliverable of the project but makes no mention of the work that goes into producing that deliverable. The PBS does not mention design work, inviting tenders for gearbox manufacture, testing, assembly or a whole host of jobs that need doing but will not be visible in any way in the final product.

Engineers produce these structures so as to be reasonably sure that they have included all the bits and pieces. The end result of a PBS in engineering terms is a long list of bits and bobs which can be passed to a storesman.

A small part of such a list would look something like this:

```
suspension
      front suspension
            off-side front suspension
                  hub assembly
                        lower support assembly
                                    pin
                                    nut
                                    bolt
                                    washers (2)
                                    spacer
                        upper support assembly
                              more bits
                              and so on
```

This racing-car manufacturer kept all of this in a CAD system and the list was used for all sorts of purposes – packing spares to take to race meetings, ordering replacements and as a basis for keeping adequate stock levels.

The higher level of the PBS is often very similar to the higher levels of the work breakdown structure (WBS) but one goes down into detail of the components and the other goes down into detail of the work to be done. Guess which is which.

Let's do a PBS for a meal. Let's consider the elements and components of the meal, ignoring what we have to do to cook the meal for the moment. Later we will have a go at a WBS which will show the work that has to be done to achieve the meal. We had better make this meal into a repast otherwise it will be too trivial and we don't want to make a meal out of it. Well, actually we do.

The end deliverable is a meal. We can do the first, one-element PBS like this:

```
┌─────────────────────────────────┐
│                                 │
│         a wonderful meal        │
│                                 │
└─────────────────────────────────┘
```

Not a hugely-useful diagram but valid and legitimate, which at least separates it from a politician. Let's expand. We'll serve drinks with nuts and crisps and stuff whilst guests arrive, then a starter, a main course, a cheese course then a sweet. There will be wine. So the next shot at the PBS looks like this.

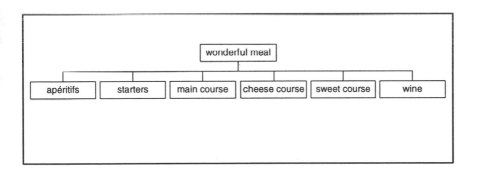

All we have so far are the main parts of the project's deliverable, perhaps the main subprojects, phases or subassemblies. The breakdown is not chronological and does not involve itself in worrying about what has to be done, it merely breaks down the objective of the project into some sensible and convenient groups. The boxes that make up this diagram are called boxes by some down-to-earth people, elements by people who are reasonably close to the ground but educated and nodes by people who want to impress. Elements and nodes are all useful terms to use when describing these kinds of diagrams. I'll use a random selection in the pages that follow.

The lines between the elements should not be called 'links' or 'dependencies' as these are terms used in PERT and this would cause confusion. There are times when your career prospects totally rely on creating confusion in which case feel free to call them anything you like.

'Connections' isn't a bad term for the lines between elements.

Before going too far with this, as with any PBS, it is a good idea to ask 'Is this the right way to start?' There are always alternatives and it makes sense to discard them with knowledge rather than in ignorance. In what other ways could we start this PBS?

We could have started at level two in the diagram with 'hot food', 'cold food' and 'drinks'. This would have brought together things that need cooking (soup, fish, crepes) and things to be served cold (snacks, cheese, ice-cream). Gin, sherry and wine would be grouped under liquid refreshment. We could have started with 'raw food' and 'prepared food'. Raw food would include the fish, vegetables from the gardens and the cheese. Prepared food would include the wine, ice cream, tinned soup and the tinned potatoes. We could have started with 'things that need doing days before' and 'instant foods'.

I'm getting hungry after all this talk of food so I hope you've got the idea that there are many ways to start a PBS and it is worth thinking them through. Think of some more whilst I get a sandwich...

Mmm. That's better. Of course if you do decide later on that your first split could be better you will thank that clever man at 3M who made it possible to rearrange those post-it stickers in a few minutes.

Let's go a stage further.

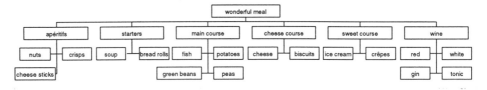

We've gone down another level thinking purely about the components of this meal. This is useful because we can do a number of things with what we have so far. I'm not going to carry on deeper into this PBS because you should have the idea by now, so let's think about the values of this diagram and one like it for your next real project.

- In the process of preparing the diagram I made a number of decisions about the exact menu. I've now got a clearer idea of what this meal is about.
- We've got a list of the things we'll need – if we did the next level we would probably have a comprehensive shopping list.
- We have grouped things into courses (phases, subprojects) so we have neat packages.
- We can show anyone who is interested what the meal is planned to contain – we have a means of communication.
- We have reduced the chances of missing something out by accident.
- We've got a structured system to which we can apply a numeric coding system (see below).
- If we had a team of people we could allocate these high-level jobs to them.

This last one is very interesting and something that happens with a diagram. We could set someone in charge of the wines, the starters and the cheese. Allocating someone does not mean that they will do whatever is necessary but that they will make sure it is done. One way of making sure things are done is to do them yourself. Often there are too many things for you to do so you allocate or delegate some jobs. Equally often you spend more time delegating the work than it would have taken you to do it yourself.

## Numbering systems

If we allocated numbers to the diagram against each box, element or node we would have a numbering system a little like that used in a library. Let's call the project '1' which is not very imaginative but workable. The next level would be numbered 1.1, 1.2, 1.3 and so on. I get the 'cheese course' element to be 1.4.

Down again, as the passengers on the sinking rescue vessel said, and we can give the nodes below 'cheese course' numbers like this: 1.4.1 and 1.4.2. I think 'tonic' should be 1.6.4. If this is even slightly confusing I'll number one of the WBS charts below to hammer home the point.

That's enough about PBS diagrams. What we have yet to do is to think about the work involved to get this meal on the table and eaten. Let's move on to work-breakdown structures (WBS)

## Work-breakdown structures (WBS)

WBS diagrams indicate the work that has to be done to achieve the project's objectives. We need to think now about effort and contribution and things that take time. We need to think about warming up the cooker, which will contribute to many items on the menu but which has yet to appear in the plan.

Once again there are many ways of starting a WBS but we could use the PBS first level as in this case it relates to the phases of the project neatly. Therefore in this case, the first two levels of the WBS look exactly the same as the PBS.

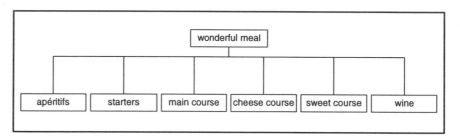

However, in a minute we'll get to look at the jobs that need doing and make up a WBS from this starting point. Before going off half-cocked we had better ask that question again – is this the right way to start? As we are now talking about workloads should we be think about 'kitchen work', 'dining room work', 'lounge work'? How about 'pre-preparation', 'cooking', 'serving' and 'eating'? Once again think about your own projects and come up with a few different approaches to start this process.

Almost certainly you will find duplication within a WBS. If you start with a component-related first level you will find yourself repeating work items later on – you'll be cooking this, that and the other. You'll have to serve many things. If you start with job groupings you will find component duplications at lower levels. Never mind, no one, except the people who make TV ads, said the world was going to be perfect.

For the moment I'll stick with the first and second levels of the PBS as a start point for the WBS. I'm only going to expand a few elements for a couple of levels so that you get the idea. Otherwise it will be so small as to be unreadable.

You should see here how we are dealing with jobs that need doing. You can also see here the numbering system at work. If I say that '1.3.4.5 cook peas' is a 'work package' and that it sounds very much like an activity in a project plan would you believe me? I hope so 'cause it's all true.

Did you notice a different but entirely valid way of laying out the diagram? The layout has been chosen to show you a different format. There is nothing special about the different layouts – this layout seemed to suit this diagram rather well.

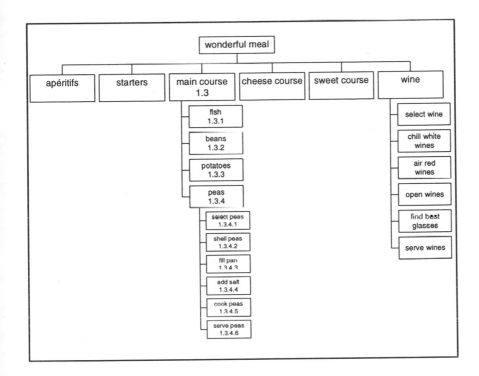

I took item number 1.3 'main course' and broke that down into its constituents. In this case I stayed with physical components at the second level, for example '1.3.1. fish' is the node dealing with the fishy part of the meal. I've taken '1.3.4 peas' as an example and broken it down further.

At this stage I have switched to tasks, otherwise known as activities or jobs. For example, '1.3.4.6 serve peas' is clearly a job that needs to be done. This is a time-consuming job that someone will have to take responsibility for and which will take some resources to achieve. Such an item would never appear on a PBS as it does not form any part of the deliverable – it is only a means to the end.

Most WBS diagrams start at a high level with elements of the project and switch into jobs and tasks later on, lower down. Most WBS diagrams have nodes in the higher levels containing nouns ('phase 4', 'tower block', 'design phase') and lower down switch into verbs and nouns ('cook peas', 'build tower block', 'execute design'). This is an important difference. Lower down the chart we get to elements that are essentially tasks and these involve doing something using what school teachers call 'doing words'. The upper levels of a WBS are 'suitcases' that form convenient containers, inside which you can group related packages of work. The lower levels represent that work being done.

Elements in a WBS are sometimes called 'cost centres'. A cost centre is useful pot of money which can be spent to achieve some part of the project. Many people relate their budgets to their WBS diagrams so that cost centres appear in both. The next logical step is to make someone responsible for each cost centre. I'll stretch my wonderful meal example a little to say that a senior project manager might be responsible for 'main course' and she might have four project managers reporting to her, each responsible for the next four cost centres – 'fish', 'potatoes', 'peas' and 'beans'.

Everyone involved in any organization knows that whenever work is done, wages are paid. Also material invoices are delayed for 60 days before losing the invoice, asking for a new one, waiting for the next cheque run, losing the MD on holiday so he can't sign the cheques, waiting for the next cheque run, sending out the wrong cheque second class, cancelling the deliberately issued wrong cheque and eventually, after months of delays, paid – all before the cost can be attributed to the appropriate cost centre.

If this was all done fairly the project managers would have an easy job of monitoring their costs for their cost centres. If all four of our hypothetical project managers keep their costs under control, their senior project-management boss will be keeping her costs under control and so on up the line.

Regrettably this is not always done fairly. Naughty and unfriendly contractors often insist on being paid for what they have actually done totally ignoring the budget for the cost package. The fact that the budget allowed £9 000 and their quote was for £12 000 is a fact they seem oblivious to.

And, to make matters worse, there is a breed of internal project or departmental people who will try hard to find any cost centre onto which they can offload their costs. They start with an invoice which will put their own job over budget. Perhaps some material had to be reordered due to an error in the specification. These underhand project managers will search around for any unguarded cost centre so that they dump the costs onto your project, making you go over budget and safeguarding their own behinds.

You can recognize this type of project manager because they are all carbon-based life forms living and working on the planet Earth. Either join them or watch it or both. Guard your cost centres with your life. If your costs go over the top and you get pulled up in front of the programme director, reasoning that your project did not want four tons of Barbie dolls and they must have been incorrectly assigned to your job will be seen as a poor but ingenious excuse. Refusing the cost in the first case or catching it quickly and sending it back leaves the problem with its maker.

Another term for a cost centre is a work package. Some WBS diagrams leap straight into the jobs level – how about the typical development phases of a software system:

- define requirements;
- write functional specification;
- write detailed specification;
- write code;
- test;
- install;
- maintain.

This is a first-level breakdown and clearly relates to work rather than elements of the deliverable or product. Similarly, a shopping-centre construction project could start off with the following items at the second level:

- shopping mall;
- car parking;
- office block;
- power house;
- external landscaping.

Under each of these there would be tasks like 'concept design', 'detailed design', 'construction' and 'letting'.

You could, with equally validity, start off with:

- land purchase;
- concept design;
- detailed design;
- construction;
- letting;
- opening.

and have at the lower levels under each of these the list of parts of the project. Under 'concept design' there will be 'shopping mall', 'car park', 'office block', 'power house' and 'landscaping'. I'm trying to get you to see the many ways that exist to do these things. If I show you lots of routes, you can probably find the one that is right for you.

WBS diagrams have the advantages of:

- helping the project team to think about the work;
- getting agreement about the work between the parties;
- assurance as the team understand what they are supposed to be doing;
- reducing the risk of overlooking some important job;
- listing the jobs to be done;
- identifying the deliverables of the project;
- making the allocation of workload simpler;
- relating to cost control 'cost centres';
- communicating the workload;
- providing a numbering system;
- listing most of the tasks for a project plan.

WBS diagrams ignore completely the logical flow of work. Almost whenever I have watched a keen and enthusiastic group of people working to create a WBS I have noticed a few very consistent symptoms.

- If there is one woman in the team she will be given the job of writing out the post-it notes unless she is very firm.
- The group will discover many things about their project that were previously unknown.
- They take it in turns to contribute and rest. At any one time, some will be thoughtful, some rushing around sticking up stickers and some will be resting.
- They will have fun – they will regard the exercise as 'better than work'.
- They will try to represent the logical flow of work from left to right of the diagram.

Groups will nearly always say things like 'But we can't install the test rig until the base is ready' or 'How can you lift the drill head before the gantry is in place?' This talk about the logical flow of the work is quite productive as such questions do need to be addressed. It is important to figure out the sequence of the work eventually. However the WBS preparation session is neither the right time nor the right place. The WBS is about finding all the tasks and grouping them sensibly into manageable chunks or phases. Critical-path diagrams and PERT charts are about linking those tasks together into a logical sequence.

I usually let a group preparing a WBS go on about logic for a little while as long as it seems to help the unearthing of the jobs that need to be done. Then I stop them and say that logic comes later, just worry about content for the moment. They always continue but look a little dissatisfied with this.

I guess some discussion about how things should be done, when things should be done and the flow of logic can be helpful but you must remember what the WBS diagram is meant to be.

## WBS diagrams and outliners

Many project-management systems offer the planner an outliner and the concept of an outliner is very similar to a WBS. Here is a small plan for the same 'wonderful meal' project showing the same structure in an outline format. Tasks are grouped under headings. The headings are bold and a slightly larger character size and these headings are the same as the elements in the WBS that collect tasks together. The tasks are indented to show how they fall under a heading. Headings can fall under headings. The highest level in this project plan is 'wonderful meal'

The numbering is the same so that, if you are feeling enthusiastic, you can check the barchart tasks against the WBS elements.

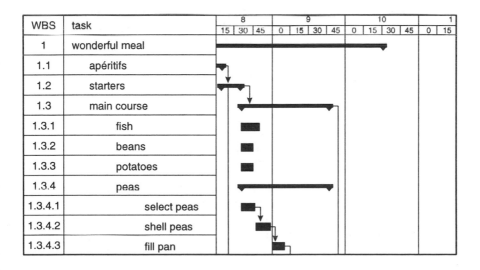

| WBS | task | 8 | | | 9 | | | | 10 | | | | 1 | |
|-----|------|---|---|---|---|---|---|---|---|---|---|---|---|---|
| | | 15 | 30 | 45 | 0 | 15 | 30 | 45 | 0 | 15 | 30 | 45 | 0 | 15 |
| 1 | wonderful meal | | | | | | | | | | | | | |
| 1.1 | apéritifs | | | | | | | | | | | | | |
| 1.2 | starters | | | | | | | | | | | | | |
| 1.3 | main course | | | | | | | | | | | | | |
| 1.3.1 | fish | | | | | | | | | | | | | |
| 1.3.2 | beans | | | | | | | | | | | | | |
| 1.3.3 | potatoes | | | | | | | | | | | | | |
| 1.3.4 | peas | | | | | | | | | | | | | |
| 1.3.4.1 | select peas | | | | | | | | | | | | | |
| 1.3.4.2 | shell peas | | | | | | | | | | | | | |
| 1.3.4.3 | fill pan | | | | | | | | | | | | | |

The tasks that have greyish bars against them are normal tasks and each of these involves doing something. Every one has a verb in its description. The solid bars with arrows drooping down at both ends like an art nouveau Zapatta moustache are 'headings'. These headings collect together a group of tasks. The word 'peas' in bold type is item 1.3.4 and is a heading encompassing the six tasks involved with preparing these stupid peas.

> Sorry, I'm a little peased off today

The bent thin lines with arrow heads are the logical links between the tasks. For example the arrow between task '1.3.4.1. select peas' and '1.3.4.2. shell peas' shows that you cannot shell the peas until they have been selected. The planning software has analysed the plan acknowledging all of these logical connections and produced the barchart displayed here.

Now that a time element has been brought into the picture with a little simple PERT logic, the tasks are laid in time across the horizontal timescale. The heading does not take any time itself but gets a bar that encompasses the overall timescale of all the tasks within it. Therefore the bar against 'peas' covers the time from the start of the earliest task to the end of the last task, which happens to be 'serve peas'.

These bars roll up so that on the highest line of the barchart is an overall cost centre or heading called 'wonderful meal' and the bar against this heading indicates the whole project and gives the total project duration – two and a half hours. Outliners, therefore, are a tabular method of drawing a structured diagram.

To achieve a barchart showing the tasks in a sensible sequence and spread out over time someone has gone through the process of adding in the logical links that show how tasks depend on each other.

One of the companies who use WBS fully is Westland Helicopters in Yeovil. They use the WBS diagram to demonstrate the elements within a modern helicopter, which I think is a fascinating machine. The following case study is extracted from *Project Manager Today* and I am grateful to PMT and Westland for their help in this case study.

## Case study: Westland Helicopters

Have you ever sat in a pub's garden, sipped on a pint of the local brew and watched your company's products frolic around over your head?

You may have sat in a pub garden and have probably sipped many a pint and the net result may have had you staring skywards from a prone position but unless your employer makes kites, aeroplanes, stars or clouds the third element is likely to have eluded you. Not so for Jim Callaghan, who works for Westland Helicopters in Yeovil, Somerset.

Jim's route to his current berth as Project Services Manager at Westland's started on board RAF fixed-wing craft. As a navigator on Canberra, HS125 and Vulcan aircraft, Jim spent his time learning how to navigate by the stars whilst flying around the world. Invalided out of the service he got a job as a PERT engineer. 'I had done a one-day course on PERT with the RAF', explains Jim, 'which gave me minimal qualifications for the job.' Fortunately ML Aviation, who make ejector units and other munitions carrier systems, were looking for personalities, not expertise.

'I came to Westland to do a PM job in 1983 but, after a few months, I became Material Control Manager to introduce a systems approach to procurement of metal', recalls Jim. This was followed by a year or so as Industrial Engineering Manager, after which he got into planning.

We talked about helicopters which gave me two surprises.

Surprise 1: you don't just buy a helicopter. I suppose that you are a normal non pools winning project-management person whose chances of buying a helicopter are smaller than your chances of getting a Maxwell pension. If you have been looking through helicopter brochures, reading *What Chopper* and gazing longingly at the displays at your local air show, you might be concerned to know that the actual purchase cost of a military whirlybird is around 12% of what you get committed to. You, as a potential buyer, would probably be a government and would have to face up to pilots, pilot training and a pilot training school, maintenance crews, maintenance engineer training, a comprehensive spares package, ro provision spares package, manpower – on-site training, support, maintenance, fuel and fuel-dump equipment, post design services – repairs and overhauls, updates of publication, configuration management, air and ground manuals, technical publications and all sorts of other bits and pieces essential to the happy-go-lucky helicopter fleet owner. This list even covers buildings on large helicopter supply projects.

I am sure that you get the idea that buying a helicopter is not at all like buying a Ford Granada. For a start they don't lick your shoes at Westland. And there is no sunshine roof.

Surprise 2: a helicopter is a flying computer. You might get the idea from assorted American films that real men fly helicopters; against all odds they haul their protesting machines over dangerous territory, constantly rescuing people stranded on boats or half way up a cliff somewhere. You might believe that real men don't eat quiche and do fly helicopters. The truth is very different these days. The pilot, for it is he sitting there in his shiny helmet, is more accurately riding in a flying computer than flying a bucking bronco.

A typical helicopter costs roughly five million pounds. Manufacture of the airframe costs less than one million, plus there is a material element of less than one half million. The remainder is bought-out kit, most of which is electronic.

Westland build an airframe – the motoring equivalent of body/chassis – but buy and fit a wide range of other manufactured equipment, including engines, avionics, radar, forward looking infrared (FLIR), altimeters and other instruments, radios (uhf, vhf, hf) and most of this rig is hooked up to an industry standard 1553 databus.

We are talking here about military aircraft where a central tactical system (CTS) integrates these system via the bus.

Jim talks about helicopters like a computer person. He speaks of buses, system integration, technical manuals – 'We are systems integrators,' says Jim. The fact that his computers whiz about the sky is almost incidental, making the mini/microcomputer look simple. After all the microcomputer just sits on your desk processing information whilst the helicopter zips about the sky processing information. The bit you see and think about is the easy bit – the airframe – which equates to the light grey box that the minicomputer lurks within.

There is of course a power train. One or more bought-in engines power the machine via 'the guts' in the shape of the gearbox. This component transfers the power to the spinning rotors and shaft horsepower goes a long way to determining the performance and load capacity of the aeroplane – this is also a bit that needs a lot of maintenance.

Talking of engines, do you know about off-engine landings? It is a part of the helicopter driving test that does require nerves of steel. To qualify as a helicopter pilot you have to land the machine with the engine turned off! Readers of a nervous disposition should skip a few lines whilst I explain. A helicopter has two key controls called collective and cyclic pitch. The first controls the general pitch of the blades and therefore the lift generated, the latter increases pitch at one part of the blades' circle so that the blades dig into the air more as they pass through that part, pulling the aircraft in that direction. As you push forward on the joystick the cyclic pitch is set to the forward position, the blades dig into the air in front of the cabin and the helicopter moves forward.

So here we are on our helicopter driving test a few thousand feet in the air and our instructor turns off the engine. Ignoring the desire to urinate we set the collective pitch so that as the helicopter falls through the air the blades windmill and gather speed. At the right moment we pull the collective lever right up using all the rotor's speed to bite into the air and slow our descent. Do this at the right moment and you land smoothly on the ground with a gentle bump. Do this too late and you are still slowing down when you bust the aeroplane to pieces against the unmoving Earth. Do it too early and you lose your downward speed but then start falling once again as gravity notices that you are hanging around quietly in the air in just the way helicopters without engines don't.

By the way a no-blade landing is much less desirable. It is for this reason that the large nut on the rotor that holds the blades onto the top of the helicopter is called the Jesus nut. He is the only person likely to be of any use to you and his name is likely to spring to your lips if this nut comes off.

Westland are proud of a number of helicopter design features, a couple of which we can briefly discuss. There is for example the ingenious ACSR – Active Control of Structural Response.

In a chopper there is a huge fan over your head which, however well balanced, generates loads of vibration. Apart from being very uncomfortable the helicopter is actually trying to shake itself to pieces. ACSR measures vibration when and where it occurs and feeds a negative vibration back by tapping out of phase. This leads to an 80% reduction in vibration, which leads to more comfortable aeroplanes that work better. This is a big issue in commercial carriers where Martinis may now be shaken or stirred and is a bigger issue in defence where people get less sapped. People still get zapped by the armaments but they get less tired doing it. Similar techniques exist in acoustics and radar where you measure incoming signals and send off the opposite signal thereby cancelling the effects.

Another refinement is the blade design, which clearly affects performance. Westland hold the world helicopter speed record at just over 400 km/hr, a record set by a Lynx called G-LYNX. This was partly a result of the BERP blade ('scuse me) which is the outcome of a government-sponsored project to design and build a super blade with a constantly changing profile, good aerodynamics and paddles at the end – a British success story.

The aircraft customer has changed – 10/15 years ago you bought an aircraft plus a box of spares. These days it is a complete package including all the things I mentioned before such as spares package, guarantees of performance plus minimised stock holdings and computer models of maintenance based on anticipated rate of use. You can see that you don't build one helicopter, deliver it and get the cheque. It is much more complicated – hence the need for PM.

Jim explains, 'PM was always needed but these days it is essential – PM is a corner stone of this company.'

## Measuring the risk

PM rears its head as risk management in one area of the business.

'Historically', says Jim, 'little work was done on risk – a little surprising as these sort of projects are very large and span a long period of time. We intend to run a pilot scheme (ha ha) on one of the current major development projects to create a risk register. This will probably use an external consultancy to back up the Westland team and will define each potential risk.' Jim lists as examples; 'a change in interest rates, inflation rates, exchange rates'. Also he mentioned a change in government in the client country.

The risk register defines each risk, explains what it will mean, discusses what action could be taken to minimise the likelihood of the event occurring and what action might be taken if the event comes to pass.

Westland intend to use a market leader in risk management for two reasons which our interviewee explains. 'We are not arrogant enough to believe we cannot learn from people who do risk assessment for a living, plus their expertise in presentation of their solution has been developed over a number of years and will ensure that we finish up with a complete evaluation of the problems.'

Jim holds strong views on risk.

'In some industries the risks are hidden by enthusiastic PM teams and we want to be right up front with it so that we go into these major projects with our eyes open.

'We can talk to a new customer about a helicopter for ten years before we get the order. Add three years for delivery. Take into account support and you are talking about 25-year projects. Not that many going on at any one time, probably four or five major projects at any one time – we would not want to do this badly.

'If we pull the plug on any one project there would be a huge effect on the company and the local population as we are the major local employer.

'It is important to evaluate risk and get agreement with your senior management so that the project is entered into with understanding of the risks – the risks are shared. This is qualitative risk management – some maths can be applied to things like interest rates but changes in government do not lend themselves to a mathematical approach.'

## When is a WBS not a WBS?

Jim opened my eyes when he starting talking about work breakdown structures.

'Often a WBS is not a WBS at all, it is usually an equipment or 'product' breakdown structure. Let's take an example often used – an aeroplane. At level one we get wing left, wing right, tail plane, engines, fuselage and then the wing gets broken into flaps, skins, stringers and so on. These are really components not work that the PM team have to do.

'We at Westland verbalise it all – we try to ensure that every item has a verb – examples might be: define training plan, prepare a specification, procure castings, assemble gearbox. The lowest level is always something one person can manage.'

This type of WBS defines what you are going to do – each element has a number and description. Does this sound like a PERT chart? Yes, by Whirlybirds, it does. If you transfer the element list into a planning package (Westland use Artemis) as a task dataset and add in restraints, durations and accountability, hey presto, you have a network and a plan which is absolutely compatible with the WBS. You will need some discipline to keep them compatible as changes happen.

Of course users of project-management systems with outliners (*SuperProject*, Microsoft *Project*, etc.) should already be familiar with this concept but many others will not be.

Every item in the WBS is a task within the plan and vice versa. We are here down to job level – people book time to each job. The WBS reference number attracts people, time, materials and any other costs. The company accounts is a separate system but uses the same numbers so that you could read across from the accounts system actual spend to date to each task against each WBS element. This means that you are well on the way to an Earned Value Analysis (EVA) system.

We have Actual Cost of Work Performed. A budget is set against the network – the WBS would do just as well – this gives Budgeted Cost of Work Scheduled. Add in progress monitoring at the plan level in terms of percentage complete and EVA is sitting in the corner and working well. This is not required by Westland contracts at the moment, it is only required by the American Department of Defence at the moment.

Items in the WBS that hold other tasks within them are treated as hammocks by Artemis – they could be treated as subproject titles or headings in other packages. It should be possible to take a WBS out of the plan feeding back so that as the plan is updated the WBS can be recreated easily plus the WBS can then have responsibilities and dates printed out on it.

The point of this is management reporting by level – in a manner that is succinct enough to make the manager want to do something about it.

Jim showed me that the WBS has tended to be corrupted from what it was meant to be. It has become a component structure relating to components and not the work that has to be done.

My thanks are due to Jim Callaghan for making me think about work breakdown structures again and for an interesting day in Yeovil. Also my thanks to Dr Ned Robbins of Origin Consultants for making the introductions that made an interesting day possible. My only regret is that, as I expected, I didn't get a trip up in one of the Westland products. Any offers?

## Organizational breakdown structures (OBS)

Organizational breakdown structures (OBS), as you might have guessed from their name, break down the structure of the organization. These have variously been known as management structures, organograms and management trees.

There are three basic patterns of management structures:

1. The **tree form**: here the trunk represents the senior management of the organization, the roots represent the shareholders, the branches, and branches of branches represent the various levels of management until we reach the leaves representing the workforce.
2. The **wheel form**: here the hub represents the senior management, the spokes indicate the lines of middle management, the rim the lower management levels and the tyre is the workforce.
3. The most common form is the **mushroom form**: here they keep you in the dark and shovel manure on you twice a day.

These little definitions are included for those new to management – everyone else will have heard it. Back to work.

The typical OBS shows the formal relationship between the functions within the company. Starting with the board of directors, the tree branches out showing how the various departments are grouped under directors, senior managers and departmental managers. These people do have names but it is their functions that get plotted on the diagram.

Their names can be added, amended and replaced – names tend to change faster than functions.

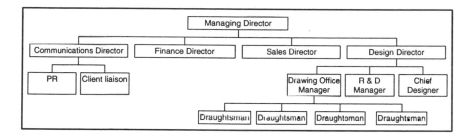

One of the reasons this sort of diagram is drawn up is to encourage you to want to climb up this tree. It is a sort of snakes and ladders board game. A reward system is put in place to tempt you to work hard and ascend the ladder of success. Just in case you can't figure out that moving higher up the ladder means greater pay cheques, symbols of wealth and success are heaped on the people in higher positions to emphasise the point. They are, by the way, heaped upon themselves by themselves.

This is why the managing director has a Jaguar, a private car parking place, a fully-carpeted office with wall-to-wall secretaries and a huge salary. It has absolutely nothing to do with being 'rewarded for the extra responsibility'.

This strategy generally fails as the motivation is misaligned. You are supposed to work hard and do everything that you can think of to the benefit of the company and therefore rise through the organization reaping the rewards.

This falls down because of the direction of this powerful motivation. There are considerable shortcuts and underhand methods of rising up through the organization and reaping the aforementioned rewards which involve doing things which are at best neutral and often downright bad news for the company as a whole. People are motivated to do what's best for them, rather than best for the company.

In addition, if you rise through the organization and have a working memory, you will remember that the predominant feeling experienced by your colleagues towards you is envy. This envy is generally translated into 'How come he is in a position like that when he couldn't organize his way out of a paper bag?' Respect is earned personally and hardly ever comes with position.

One of the other major elements in an OBS is money. Dosh, cash, loot are all supposed to increase as one moves up the organization. You are supposed to be motivated by money. Curiously whilst most people are happy to show off their lavish office, company car and other benefits, no manager talks about how much they actually earn.

I know of two fairly senior managers who were supposed to be roughly on par and agreed after talks of great length to tell each other their salaries. So worried were they that one of the two would cheat that they agreed to write down their salaries on bits of paper and swap. One bit of paper said £25 000 but the other said 'Gotcha'.

Isn't it strange that in the world of management we are keen to show off so much and are so secretive about salary. Money of course is the one perk that is very precisely measurable. Loads of work has been done to show that money does not motivate management people very much, the idea or promise of more money motivates but cash actually doesn't help much.

Many people are happy to chase a Jaguar, a secretary and wall-to-wall management trees and are happy with the idea of people being envious of them. Such people play the management-tree game for all they are worth. This seems to fill the working life of politicians and other government employees.

People not interested in playing political games should drop out or join small teams and at all costs avoid large organizations where such basic laws are taken to wild extremes. For me, small teams are beautiful.

*The Peter Principle*

This philosophy of work hard, climb the tree and get loads of dosh sits badly with technicians who happen to enjoy and be good at their trade. Start off with a great programmer who gets noticed for being good. She gets promoted to section leader and handles that well. She gets promoted each time she spends a little time doing a job well. Eventually she becomes senior project manager which is a job that involves all sorts of politics and team building which she happens not to like.

She probably gets a little training for this new job but it is woefully inadequate for her needs. She yearns to be back with those intricate programming problems which she enjoyed dealing with. As she cannot handle the job of senior project manager she does not do well and does not get further promotions. She gets stuck in the first job she can't really handle.

Now the organization has an inadequate senior project manager and has lost a great programmer. The 'climb the ladder' approach provides no system for stepping back down a few steps or even for stopping in a job you like and can do well.

Therefore management guru Mr Peter said:

> Everyone tends to rise to their own level of incompetency.

It takes a great deal of perception, strength of character and judgement to refuse a promotion. You will need these things not to turn down the promotion but to deal with all the people who feel threatened by your rejection of the philosophy they hold so dear. The list of people who feel threatened will include the person that offered you the promotion, their bosses and so on up the tree into the distance. Plus the personnel department.

I would like to create a management tree that was completely the other way up. Instead of being shaped like a right-way-up pyramid it would be an upside-down pyramid. The front line, the people that do the business, the ones that actual keep the business going would be the top line and there would be levels of supporters below that line. There would be no pressure to move down the tree and rewards would follow the company's success, not the individual's success. I don't suppose it would really work and I can't figure out an appropriate pay structure but I can dream can't I?

It is also well known that OBS diagrams do not show how the organization actually works. Studies have been done into the communication channels that really exist and how frequently these channels are used. By plotting the members of the organization and connecting the names with lines representing the frequency of contact with other names you get a picture of how the business really functions.

What do you really speak to your boss about? The answers are probably:

- Personal issues: you talk about your career, your qualifications, your training and, the really important stuff, you book your holidays. You might mention your miserably inadequate pay from time to time.
- Personal expenses: these need approving from time to time.
- Mistakes: your direct line manager, the person immediately above you in the OBS, is the one who is supposed to criticize you and warn you if you go on doing things wrong.
- Your next job: your boss, your immediate superior, is the one who should 'let you go' if your work fails to meet the required standard. Your boss's desk is the one you are supposed to aim to sit behind if you do well.
- Your next assignment: your boss should hand out assignments to you, making sure that you have the authority and skills to perform that assignment. Simultaneously your boss should take responsibility for the work you are about to do, in the same way as your boss accepted the assignment from above. The lines on an OBS show who should trust whom. Your boss should trust you and you should trust your subordinates.

Maybe your boss is technically experienced and can therefore help you to solve technical issues. This would be a happy coincidence. It is more likely that the most competent person to whom you go for technical help is in another department all together but is kind enough to help you out. It is equally likely that your respect for your boss' technical ability is low compared with Genghis Khan's politeness, which leaves you wondering how that person and that job ever came to be joined. Remember that one day someone will feel exactly that about you. That one day could easily be today.

*The politeness of Genghis Khan*

So an OBS shows whose shoes you will step into next and who you will be fighting with to get into those shoes. The OBS should show who can give you the authority to do something and, at the same time, take responsibility for the way in which you do it. It also shows who will reward you, warn you and, if necessary, sack you. The OBS also shows you which departments you can sensibly expect to win under your control in the endless corporate battle to build empires – now that is really useful.

Much research work has been done examining the structure of organizations and a few books on the subject are mentioned in the further reading section at the back of this book. There are some buzz-words which I should mention in case you need to drop them into a conversation one day.

**Span of control** refers to the number of boxes immediately below your own. Urwick, one of the great management gurus, said that no manager should supervise more than six subordinates. Look around and you probably see that most people manage between four and seven people, which shows why Mr Urwick is a guru. It is also said that the human brain can hold six plus or minus two things at once.

Management people talk about tall and thin structures being inefficient. Tall and thin means many levels and narrow spans of control. Generally learned management writers seem to recommend that most organizations need a maximum of five levels to do work. The larger the organization, the more levels are required assuming constant spans of control. The British Civil Service, renowned worldwide for its inefficiency, had nine levels and spans of control in the order of two or three.

**Staff** and **line** are terms indicating your position in the hierarchy. Staff members make recommendations to the line managers but don't actually carry out any productive work and certainly have no responsibility.

This stems back to war-speak when staff people drove cars and planned supply routes and line managers told their subordinates to shoot other people's heads off. I am unsure about the benefits of figuring out who is line and who is staff but some organizations do make a meal of this issue.

Within the world of programme management we must face up to the position of the OBS in an organization running projects. On one hand we have the functional managers, each of whom runs a department specializing in some specific area of the organization's affairs. We have design departments, prototyping, testing, installation, optics and a whole range of other specialist groups. Within each department we will probably find a hierarchy translated into a part of the organizational breakdown structure. Sitting above the functional manager the hierarchy will extend upwards towards the board of directors.

Somewhere alongside this part of the hierarchy will be the project-management team in a hierarchy of its own. Here we might find programme managers controlling project managers who themselves might have phase managers, planners and other project-related engineers. Frequently the members of these two hierarchies swap places as they temporarily switch from a functional role to a project role. In other organizations, the two management trees are fixed and as permanent as each other.

Levels of authority can vary greatly, with designers in a functional department forced to seek permission to buy a couple of new pens whilst project managers have huge budgets to wield on their own. Project managers might be able to place large orders with contractors, hire and fire staff and buy large quantities of goods without reference. Frequently projects people zip about the world leaping on and off planes and trains and staying in hotels whilst the functional groups stay at home. Some organizations, long accustomed to dealing with strict levels of authority, find it hard to adjust to people with huge expenses and a need to commit to large expenditure quickly. Some organizations find it hard to do anything quickly after years of gentle, steady and predictable business.

Finally, after this brief introduction to the world of OBS, one neat idea that I bumped into at the Rolls Royce engine-design plant in Derby. Here, a group of a couple of hundred specialists had been thrown together into a project team to design and build a new aero engine. There were huge numbers of design issues to be addressed and many simultaneous tests going on. People would set up a test on one part of the engine including the monitoring equipment and the run conditions. Meanwhile many others would be doing the same thing on other aspects of the engine's performance. A clock on the wall showed when the next test run was due and each group tried to get their test ready in time. This threw members of the many design groups into contact with other in efforts to co-operate about their widely-different interests over the costly test runs.

To make things simpler amongst the members of the team – many of whom were complete strangers to each others, the manager put the OBS up on a large wall in the design office and stuck a photo against each person's name. This made it much easier to find the fuel-line supply-team leader and to identify the person near the third window with the big drawing board.

## Responsibility assignment matrix

This little document is designed to link the WBS – the work to be done – with the OBS – those available to do the work. It is sometimes called a RACI matrix – you'll see why in a moment. It also sometimes called a **linear responsibility chart (LRC),** which is a very grand sounding name, the sort of name that gets you a PhD just for dreaming it up.

We go over now to the Blue Peter studio where our resident presenter will help you to make a RACI matrix. All you will need is a small thermonuclear explosive device (get your mother's approval first), fourteen cabbage white butterflies, a cardboard box (detergent size), some glue, three toilet-roll holders and a piece of squared paper.

Take the piece of squared paper and place it centrally on your desk. Now imagine taking the OBS and placing it on the desk the normal way up just above the squared paper. You would have all the names of the workers and their functions across the top of your squared page.

Then take the WBS, turned in sideways and put it down to the left-hand side of the squared page – this gives a list of tasks in a column down the left hand side of your squared page. You would achieve the same objective with the outliner mentioned before. Take any task, run your finger across the page and stop when your pinky is under the name of the person who is responsible for doing that task. Mark that box with an 'R'.

Every cell in the matrix is completely filled with either a blank or a letter showing some sort of involvement for that person with that work package. British Rail signalling engineers use an R, A, C or I, meaning **responsibility, accountability, consultation** (in some companies **communication**) and **information** at each point. This shows who is doing what on the project. You could develop your own system if you don't happen to like BR's. There could be one such diagram per phase of the project if it gets too big.

There you are, your very own RACI matrix which you can take to show your boss. If he doesn't like it you can use the thermonuclear device to blow him off the face of the Earth (get his permission first) simultaneously promoting yourself and getting the 1.6-litre XL car you so richly deserve.

Before going to see the boss with your RACI matrix you had better be prepared in case he asks what these letters stand for. You had also better be prepared for him to speak correctly and not end his sentence with a preposition – for what do these letter stand?

Here are some ideas but please do not take all of this too literally as many organizations use their own interpretation of these terms. Anything sensible can be used as long as the people in your organization know what the letters mean.

### Accountability

This is person who is accountable if this task goes horribly wrong. This person may or may not be actually doing anything at all towards the achievement of this task but it is this poor person who gets a slap on the wrist if it is not done correctly, on time and to budget. Only one person can be accountable.

### Responsibility

These are the doers who will actually get their hands dirty. My experience of project management is that the only times that project managers gets their hands dirty is when digging carrots, changing nappies or toner cartridges. Only the last generally happens within the office so I put all this talk of 'rolling up our sleeves' and 'getting our hands dirty' down to wishful thinking – they are wishing that they could actually do something useful.

Still, those with an 'R' in the RACI matrix are expected to do the work. There could be many people with responsibility for doing the work – it could be a team of people – and they would typically report up to the person who is accountable for their actions. On a line across the matrix there can be many 'R's but only one 'A'. It is nice to know that you can roll your 'R's across the page.

One person can be both responsible and accountable for a task and would therefore get 'A/R' in the relevant box.

### Consultation (communication)

A 'C' represents a two-way communication between the doers and someone else. Some people call it consultation. In the example below, the Chief Tester needs to be consulted about the User Tests and reckons that he will have something to contribute to the Market Researcher.

### Information

An 'I' in the box indicates keeping an eye on the task. Here the Senior Designer wishes to be kept informed about all the testing tasks so that he can evaluate their impact on the design he is accountable for (oops, for which he is accountable).

Here is one we prepared earlier:

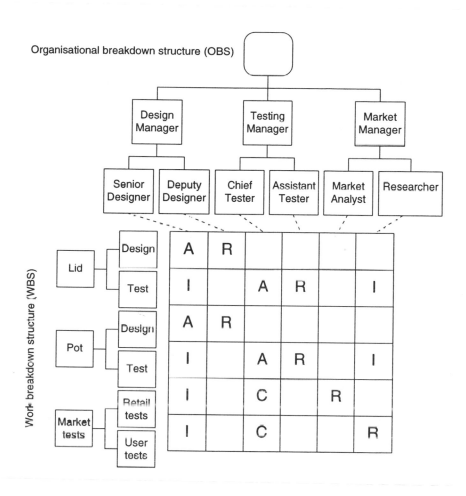

You can see that, for example, the 'testing the teapot lid' task involves four people. The Assistant Tester is responsible for carrying out the task and the Chief Tester is accountable for the test being done. Both the Senior Designer and Market Researcher wish to be kept informed of the testing process.

The values of such a diagram are that:

- Decisions are taken about who is doing what – this tends to reduce the number of times everyone in a meeting looks blankly at each other saying 'I though you were doing that.' Misunderstandings are reduced.

- The roles of each person are easily communicated so that people who want to know can know who is doing what to whom and with what.
- Extremely busy people can be spotted and checked against timescales. Overloads can be planned around in one way or the other.
- Before the thing that is brown, steamy and smelly and comes backwards out of cows hits the air-conditioning and covers everyone, the blame can be reasonably fairly allocated.

| The Isle of White Ferry of course. |

- Regrettably more rarely, praise can heaped upon the right person. In our world where so little praise abounds, any going should at least be fairly placed

As soon as the RACI matrix has been drawn and distributed, there is a danger of demarcation. Do you remember those days when unemployment was minute, jobs were easy to find and strict lines were drawn between simple and similar tasks? Some were undoubtedly sensible – an airline hostess (aka trolley dolly) might not be the right person to fly a plane. Some were ludicrous – a labourer could lift cardboard boxes but not wooden ones. People were prevented from doing things at work that they would do without a second thought in their own homes.

> You might find a RACI matrix for your family revealing.

If you value a team spirit then take care with your RACI ideas. You might drive people away from the concept of 'we're all in this together and we should all do anything we are competent to do'. You might move towards 'That's not my job', which is right and proper in some situations but can be divisive if someone refuses to photocopy the report because it 'isn't my job'.

# Multi-project planning problems

## Resource conflicts

Resource conflicts has to be top of the list when it comes to the problems of planning a series of projects. I hope by now you have begun to rely on me for the normal official educated line plus the truth – well this is another example of those two aspects.

The theory says that many projects vie for the same resources. This is clearly likely to be true and is a problem that requires addressing. 'Addressing' in this sense does not mean sending the problem a post-card showing your holiday hotel in Bonka Plenti.

There are often many projects going on, each of which will cause a demand for a skill. Tasks require design input, welding and painting. These are skills, not resources. People are resources and people have skills. Geoff and Jenni are resources and they might have skills in painting and dogwalking.

The requirement for skills can be provided in a number of ways. There may be software programmers who are the obvious people to meet the demand for programming skills. These programmers have names like Fred, Azif and Julie and these are resources. They are capable of producing work and meeting the skill requirements. Their ability to provide programming skills is probably around 100%.

You may have a few junior programmers or system analysts who do not normally write code but do have some of the skills. They may not be as efficient as the proper programmers but they will get there in the end. Perhaps they are 50% efficient at programming.

To use another example Joe, the caretaker, is 100% efficient at sweeping the floor whereas Sue, the dinner lady, is only 50% efficient at the same job. Albert, the Managing Director's efficiency at sweeping the floors is…if only, if only.

Some people record each resource's effectiveness at performing skills. This can become a political hot potato as people get upset at being graded, especially if they don't come first. It is often quite hard to measure these things and a league table may get you into more trouble than you need. I'm told in some countries this sort of grading would be illegal.

It is much easier with machines. There is no reason why you shouldn't plan to use skills like 'milling' and satisfy that need with a milling machine. You might have three milling machines each of which has an efficiency at performing milling operations. The machines won't mind being graded into a league table.

There are ingenious resources and skills that people use: 'lecture rooms' and 'space' being just two.

The conflicts arise because more than one project manager has planned more than one project containing more than one task to occur simultaneously. The net result of these tasks is that the resources available cannot satisfy the demand for the specific skill. For example, there just aren't going to be enough painters to paint all those components in August.

The bad news is that you need more painting skills than you have available to do the work. The good news is, because you have been a good planner and planned the work in appropriate detail, you have time to sort out the problem. If your role is a planning role then, once the problem is identified, you are in the clear. You've done your bit and you pass the problem up the management hierarchy to the referee, umpire or senior manager who can decide who gets what. If you are that umpire it is time to don the thinking hat and sort out this problemette.

> What is the difference between a problem and a crisis?
>
> Time to sort it out.

When your management have decided who is to get which resources you should be told because this will probably change the plan. Perhaps someone's project has to be delayed to await the resources coming free from some other work. This needs to be shown in the next plan update so that everyone knows what is supposed to be happening.

I am rashly assuming that a system of planning exists that predicts the demand for skills and resources and that predicts shortfalls. I am also assuming that there is a mechanism for discovering future problems and resolving them. If you are in programme management and don't have any of these things, it is time to get organized.

Human nature plays a significant role in this resource-allocation business and people lie a great deal. The sort of people that you want to push, shove and heave your projects along are going to be human bull-dozers. You want keen, enthusiastic people who have a desire to get the project done. The sort of people that don't mind taking a bit of a risk in the interests of getting on with the job.

Such people will have found over a period of years that one system generally works well when it comes to grabbing resources. They will have noticed that shouting loudly is one very effective strategy. Nagging is another. Barefaced lying to exaggerate the importance of the particular project is another. Bribery, favours and a thousand other tricks are employed by the highly-motivated project managers to get their projects allocated with the resources they need.

Another neat trick is to lie about task timetables. If you say 'We want this testing doing from 14 to 21 February' you might well be told by the testing department manager: 'Impossible. We've got the new Glando model to test then and we've got three guys going off on some silly training course in programme management so you've got no chance.'

You jump up and down, throw you hat on the floor, claim to know the MD's wife personally, threaten to tell the MD about the design chief's illicit habits, threaten to take the matter to the boardroom and eventually settle for the 25–29 February. Of course this is exactly what you wanted in the first place and thank you, that will do nicely.

Yes, the truth is that people lie. And that's the truth. With the best possible intentions and in the best possible taste these project managers will do anything, including lie their back teeth off, to get their project out on time.

The better the programme-management system in place the less this kind of thing is needed. In those organizations with little planning and no methodology they simply rely on this kind of cut and thrust to get the work done.

It collapses frequently as people are constantly letting each other down with no notice at all. Design packages do not arrive for manu-facturing causing workshops to be bereft of work one week and flooded the next. Promises to do work are broken by functional managers as their resources, previously allocated to you, are spirited away onto another project with a louder project manager. This means the sub-contractor who is due next week will not be able to start doing anything more useful than submitting his first bill which will say:

To standing around all week, as per schedule, with nothing to do, due to long delays before we arrived
££££Lots & Lots         Plus VAT.

Would you like to send that bill up to your boss for approval?

There is also a trend to take life easy, ask when some resources are going to be available and plan around that.

These are all problems that surround the mystic art of resource allocation. Programme management offers a structured and well-managed approach to solving these problems where the aims and objectives of the organization as a whole can be balanced against the needs of the projects and the availability of the resources. It is up to you and your company to choose: programme management or cut and thrust.

## Responsibility and authority: the project manager's trap

I've got a bit of a bee in my bonnet about authority and responsibility and this little striped monster keeps buzzing around my hat because I have seen so many budding project managers suffer over this issue. The basic argument is as simple as it is elegant:

Someone gives you the authority to do whatever is needed to achieve something and you take responsibility for getting it done.

This is a reasonable deal and one that is entered into by most managers and subordinates regularly. It is the lack of such a deal that causes the problems. I have seen many otherwise promising managers plucked out of their nicely-moving careers to handle some prestigious project and finish up with their career dead in the water.

This is what happens. Let's take a young enthusiastic person who performs a useful and laudable function within the organization. She has moved up the organization quickly, been on training courses and is well thought of. She is a bit of a flyer and has been noticed by the powers upstairs.

The organization runs on an even keel and has no need for, and no knowledge of, project management. Along comes a project which really doesn't fit into the company's way of doing business. It may be to introduce some new products, open up a new warehouse or whatever – what it is, is not important. The potentially good news is that the MD reads the Torygraph and there was a piece in the business news about project management. He decides this new project needs some project management and picks out our rising star to do the job.

She realizes many things right away. She is going to be dealing with the senior managers and directors so this is a chance to shine in front of important people. Good. Her current knowledge of project management is low. It is second only to her knowledge of Estonian Aviaries. Bad. She is not proud and is ready to learn. Good.

Two goods and one bad, the balance is right so off she goes to learn what project management is all about. She goes on a three-day training course and reads *Project Management Demystified* and gets the idea quickly.

> Sorry, couldn't resist the plug.

On the training course she meets a charming man who turns out to be Estonian and who owns a aviary. He takes her home but they have a minor car accident on the mud in front of one of the cages. She kills two birds with one Estonian.

She assiduously plans the project, learns what has to done, defines the objectives, produces plans and takes responsibility for bringing the job in on time.

Tenders come in from the builder and she decides to place the order with the cheapest and get the building going when the purchasing officer interferes. Said purchasing officer comes from the school which teaches 'If a thing's worth doing, it's worth doing right.' This is fine if you understand what 'right' means. He interprets 'right' as being carefully checked and insists on passing the form of contract by a solicitor and hiring a quantity surveyor for expert advice.

'Right' to the buyer means 'so that I won't get into trouble'.

Whilst all these checks and double checks go on the project manager gets more and more frustrated. Eventually the builder decides this is too much hassle for him bearing in mind it is only a small renovation to a warehouse building so he withdraws his bid. The whole process has to start over with the second lowest bid.

The project manager knows that 'right' means a judicial balance of good price, low risk and speed. The purchasing officer is not interested in speed and thinks that 'fast' means 'this year'. All of his working career he has negotiated contracts for five years' supply of teddy bears at a snail's pace. Small print is his middle name.

The project starts to run behind schedule and other work suffers. Our heroine's reputation starts to sink in front of all those important people she so wanted to impress. The project does eventually get finished but long before its end she has had all the stuffing knocked out of her due to long periods banging her head against a brick wall. It ends way behind schedule and way, way over budget. She loses. There is no happy ending. The buyer says one of the worst things that any employee can say: 'It wasn't my fault.'

She had the responsibility for the project but not the authority to do things to make the project into a success. She had to break out of the company's normal way of working to respond to the needs of the project and yet she had no authority to do so. She bumped straight into a bit of bureaucracy which stopped her high-speed project in its tracks. She tried to explain but they didn't listen – as far as the board goes, she just made excuses.

So next time you are given a project and accept responsibility, make sure you have the authority to make it happen. The example above talks about an organization not used to projects but the lack of authority could come from any direction. You might need the signature from the MD to place orders over £10 000 but you can only see him on Tuesday evenings. This is not uncommon in organizations where no one places an order for anything apart from some new pencils. Placing an order takes at least one week plus tendering and evaluation time.

You are unlikely to run a good 100 metres time with your hands tied so make sure you get the authority that matches the responsibility you are taking on. Then you'll have fewer excuses but a greater chance of not falling into the project-management trap.

## Consistency

A common misconception is that programme management is actually about bringing together many single-project plans. The people who put these ideas about are generally those who sell a project-management software package that can bring together many single project plans.

If you take the trouble to read the section on 'Project-management tools in the programme management environment' you will see that there is, or can be, a great deal more to it than this. However, the idea of bringing plans together is a foundation to programme planning – the planning of multiple projects.

Some people call this roll-up, some talk about merging plans together and still others speak of linking plans. This is generally known as consolidating and deconsolidating the schedule. Consolidation is about collecting and merging the plans. Deconsolidation is about splitting the plans back out again once they have been modified in light of the problems highlighted.

The idea is that many people create many plans for many projects and, from time to time, submit their plans to a central planning function.

Within this function the many plans are brought together in some way and the total workload viewed. Software packages are very often sold on their consolidation abilities but tend to ignore the deconsolidation side of things.

The total demand made by all projects on types of resources can be viewed. You can see one histogram showing all the programmers working on all projects. As this will normally show that you need three times as many programmers as you have working for the organization, you have achieved a great deal. At the end of the ensuing discussions some projects will be rescheduled, some contractors hired in and other steps taken to arrive at a workable plan.

As you normally cannot deconsolidate the big plan back into its many little plans, this new and workable plan needs to be communicated back to those many project managers so that they can bring their plans into line with the master plan.

The MD can do naughty things with this database of all the projects. He can list them all in order of greatest lateness or overspend. He can list all projects and demand a status report from each project manager. The head of design could list all design tasks by getting the multi-project database to search through all projects for design tasks. These are very admirable ideas.

But to achieve any of this consolidation requires considerable consistency. All the project managers are going to have to agree on some conformity to their planning. Let me mention a few areas where consistency counts – you can't stop me, can you?

You don't have to read this bit if it doesn't apply to you.

### Calendars

Calendars define when the company, the project and each resource is available to work. Typically a calendar states that Jenni is available 9:00 to 5:00 with an hour for lunch five days each week. She takes all Saturdays and Sundays off as well as Bank Holidays and her birthday.

Software packages get upset and stamp their little procedures angrily if the various plans have different calendars from each other and the multi-project workload itself.

### Task names

If the MD is going to search for particular types of tasks he needs to know that everyone is using a consistent task-naming and numbering form. To a dimbo computer 'Design Nosecone' is completely different to 'Nosecone Design' which is also different to 'Nose Cone Design'.

This applies to milestones equally, which are the sort of thing that an MD is likely to look at. Inconsistent task or milestone names may mean that the MD misses out your project and does not discover that you are setting records for lateness. You wouldn't want that to happen, would you?

*Resource names*

A programmer is totally different if seen through a computer's eyes to a 'programmer/analyst'. Ms J. Buchet is another person when read next to Jane Buchet. If different project plans contain different resource names the software will add them both up into separate groups not realizing that they refer to the same person or trade. Then, when you pull up a histogram of Ms J. Buchet's workload you see that everything is fine and she is nearly but not quite overloaded. What you don't see is the other histogram which is also nearly but not quite overloaded but which shows work to done by the same person under a slightly different name. If you realize and add these two together you see the true position of this single resource.

*Cost centres*

It is common for people to add up the costs associated with each project and once again, using consistent category names reduces the risk of getting the cost monitoring wrong.

*Executive information systems (EIS)*

These sound grand and very high tech but are really quite boring. In our terms an EIS refers to the idea of looking at many projects, selecting one and digging into it to find out what is going on. This is called 'drilling down' which is why it is so boring.

You have the total corporate workload to start with, you list the projects in a given order, perhaps greatest slippage first, and see a single bar for each project, including the project with the greatest slippage, on your executive screen which nestles alongside your executive Newton's Cradle which has your initials carved in every ball.

You select this single bar and expand it to display the phase level – six or eight bars summarizing the project. You see that the design phase is well behind so you select the 'design phase' bar and expand it to display the detailed plan. You eventually find that the project was delayed whilst the design team worked on a design for a Newton's Cradle with initials on every ball.

You have a little lie down and ponder the benefits of an EIS system. One thing you are glad of – everyone uses the same summary bars to give an overview of each project. If they didn't do that your task of locating problems would be a touch harder.

There could easily be a huge volume of data. If you plan 100 projects and each one as 30 tasks you are already talking about 3 000 tasks. This is a big plan and it will be hard to find your way through this mass of data. It will not fit on even a 17-inch screen and it will take pages to print.

You need to be able to select tasks that pass some certain criteria and deal with those before passing onto the next bunch. You might hear someone say 'Let's call up all the engineering design tasks and deal with those before looking at the optical workload.'

You can only select tasks of a like kind if there is some suitable coding system in use. You might use task numbering, task names or task descriptions to help you find the group of tasks you seek. Some software packages, rather confusingly, provide this feature under the label of a work-breakdown structure or a WBS, which is the wrong name for a useful feature. What you actually get is a structured field where you can imbed codes showing what kind of work it is.

Alongside each task you get a space of twelve or so characters to code your tasks to aid in searching and grouping, such as: ENG 95 Design 1020. This might mean this is an ENGineering task taking place in 1995. It involves the Design department and it is part of job number 1020.

You can dream up your own coding systems if you think this will help. It pays to spend a little time early on figuring out what this coding system will do and how it will be used before designing the system and telling everybody about it. Whatever your systems for coding, task naming and numbering are, they will have to be consistent. There are some software tools and a couple of techniques to help deal with consistency issues.

There are a couple of software tools around that will prepare a project plan for you. *Project Bridge Modeller* from ABTI and *Project Guide* from Symantec (there is a product and vendor list in *Project Management Demystified*) both offer an automated approach to plan construction. They work like this:

Someone, somewhere in the organization, let's call him the planning guru, builds a model of the typical project undertaken by the organization. The plan will include phases and checks that must be adhered to (e.g., board approval before commencement of manufacturing) and all the activities that make up the project. The durations of the tasks and the resources that are required to perform them are also calculated.

This happens only very occasionally. Once set up this is turned into a tool which is used by the project manager when she gets a new project. She fires up the plan-creation software which starts by asking a few basic questions. What is the name of this new project? How many shelves has it? How many forms need generating? Does it need a design phase or is the design being done by others?

All these questions were planted by the planning guru to extract key information from the project manager about this specific project. Once all the questions are answered the plan-building software chunters off and generates a project plan with durations, resources, phases and milestones all by itself. This plan now fits with the company's methodology and is consistent in the terms described above.

There is another way. You get the planning guru to create a typical plan and store it under the name 'TYPICAL' or 'STANDARD'. Whenever you want a new plan you pull in TYPICAL, modify it to suit the specific project and save it away under its new name.

Yet another approach is for you to become the planning guru yourself.

## Speed

'Speed Kills' shout the safe-driving campaign posters. But slowness kills in programme management. There is a closed loop around which the programme planner travels and it is going to take some time to run around the loop to produce up-to-date project plans. The process goes like this:

1.  Get the latest data about work in hand (see later).
2.  Get the latest update on new jobs.
3.  Get the position of scarce resources.
4.  Check up on people's holidays and training courses.
5.  Update all the data in the various plans.
6.  Process the plan.
7.  Distribute the information.

This process is vital. You are doomed to failure if you do not regularly monitor and update your programme plan. You are simply wasting your time. Actually it is worse than that. If there is a plan stuck up on people's walls telling everyone what is supposed to be going on, you cannot blame people who try to stick to it. If someone goes out and buys an expensive piece of plant and starts work in accordance with the schedule but is hampered because of some previous delay, they can reasonably say 'It was on the plan'. Out-of-date plans are more dangerous than no plan at all.

To keep up to date with the changing plan of work you need to set up a system for monitoring what is going on, keeping abreast of changes in priorities, workload and strategies and producing up-to-date plans.

This process can take anything from a few minutes to a few days. If your plan takes a few days to get updated and stuck back up on people's walls it will be always be a few days old. There is a very good chance that it will be well out of date the moment it gets issued. It will not be well respected and not used. Your role will be questioned. Time to scan the classified ads.

You do need to set up a system where the turn-around is fast. Buying software that is very fast clearly helps but it only helps with step 6 above. Neat and efficient data-entry methods shorten step 5 and quick printing shortens step 7. The rest is down to the organization and the way it works and this is where you need to really concentrate.

Speed of feedback is a problem. If you plan all the work including training, holidays and non-project work but don't feed back actual progress through some efficient system the plan becomes worse than useless – it becomes dangerous.

## Timesheets

If project plans are going to be anything like useful they must reflect reality. A plan is an external model of what you think is going to go on, it is a model of a future. It may not be the model of your future, it is much more likely to be a model of a future that takes place on a distant and nearly-parallel universe where the electronic household gadgets break down one day before the guarantee runs out.

> In our universe household electronic gadgets break down one day after the guarantee ran out.

Your plan may have been spot on last week but since then three more jobs have been won, one has been cancelled, the design department had to close for half a day due to a bomb scare and a whole range of other bits of progress have occurred.

Most of these bits of progress have been achieved by sensible, intelligent people getting on with any work they can get on with. The list of actual work done will bear only a passing resemblance to the work planned for the same period so you had better update your plan in light of these actual achievements.

One way to do this is via a timesheet system. Timesheets are a great way of finding out what is going on. Setting up a system for a regular feedback to the project managers or their planners helps to keep planning in the centre of things and plays a useful and positive role.

There are some problems:

Timesheets are supposed to be a quick, efficient method of collecting data on actual work done. This is done for a number of reasons. Staff can be paid, people can be checked up on, clients can be invoiced and you can update your plans, to name but four.

Actually timesheets are some of the greatest works of fiction of our time. Many timesheets ask completely the wrong questions and collect inappropriate and misleading information. Many members of staff lie their teeth off through the medium of the timesheet in an attempt to cover for the time they spent drinking coffee and chatting up members of the sexual group they happen to fancy.

Timesheets ask what time has been spent on a task. For example, against the task name 'Design Nosecone' Designer Danny enters 40 hours in the 'Time Spent This Period' column.

What progress do you report on that task? Not forty hours as Danny has overlooked the two or three hours each day he spends in meetings, on the lavatory, reading technical journals (*Nosecone Today*), reading technical journals (*Private Eye*) and backing up his computer.

Most people actually produce project-related work for about 50–70% of their day. It is entirely possible that they work hard and do a lot of useful other things, spending no time reading *Private Eye* and keeping up with recent developments in their field on the train to work. Nevertheless, there is an overhead of time that does not directly relate to the project and its tasks.

Danny also took a look at a new job he may be doing next month. He spent a few hours giving the new job a onceover to estimate how long it would take to do the design. Very useful but non-productive as far as this task goes.

Regrettably, we find later, the 25 hours' time he actually put into the nosecone design task was based on an out-of-date specification which has been scrapped. No one told Danny. Net output for this task = zero hours. When he finds out that this has happened, he is likely to go off into a blue sulk for a few hours and the design manager will have to do her damnedest to cheer him up. This could take a couple of hours, so the true progress this week was minus two hours.

Danny reports being 85% complete this week mainly because he reported being 75% last week and progress should go up week on week. Slightly extreme, you might say, but it happens. The big question that the timesheet should have asked is:

How long is left to do on this task?

This makes no assumptions about the accuracy of the original estimate for the task's duration nor about Danny's production rate and interruptions. It makes people think about the work that is left and allows for changes to affect the time that remains on this task: the remaining duration.

So timesheets can be a great help to a programme plan. They need watching carefully. There are a number of PC-based timesheet programs on the market which can be trained to link into project-management software and in such a system lies the basis of a true programme-planning system.

## Downtime

There is inevitably a background workload plus holidays and training time which has to be dealt with somehow in a programme-planning system. Here are some time-consuming operations that do not relate to specific tasks in the specific projects but do absorb some of the time available from your precious resources:

- training courses;
- internal non-project-related meetings;
- holidays;
- travelling time;
- union meetings;
- filling in timesheets;
- talking to project planners and managers;
- reading books about programme management;
- regular background work – user support, filing, backing up computer data;
- chatting up the blonde in the corner;
- reading the *Sunday Sport, Hello!*, or *Which?* magazine.

They all take time. There are a few ways to deal with these demands on your resources' time:

### Ignore the whole thing

Generally a bad plan. You might plan for each person to do 40 hours each week for the foolish reason that they get paid for 40 hours per week. Everything will appear to be behind schedule and everyone will blame everyone else for your error.

### Plan on a realistic 30 hours per week

Much better. You could produce data or calculate for yourself the actual 'project productive' time available each week. Then you use this as a reasonable estimate of the available time each person has. This information is usually a part of the working calendar for the resource.

You can avoid upsetting people by selecting your words carefully here. If you baldly state 'Our resources work 30 hours per week' you will cause much upset between the resources and the management that pay them. The management might feel that people are paid to work 40 hours each week and they should work 40 hours each week. At least.

To avoid this kind of upset you might imitate a management consultant and use some confusing and unhelpful but technical-sounding terminology. Here are some suggestions:

- available product progress time
- effective task progress time
- availability for project work
- direct project progress time

These are the sort of terms that sound grand and technical and that soften the blow. They do not infer that the rest of the time is wasted, just that it doesn't go directly into any project. You might be able to calculate the realistic amount of time you can expect from each resource on a scientific basis.

*Allow for downtime in productive time*

In this strategy you stick with the standard 40-hour week and plan and monitor against it. When you create tasks and assign people to those tasks you allow for their downtime in the durations you estimate. Some organizations use a set of constants which show how long things take to do allowing for the non-project downtime. For example, bricklayers lay around 80 bricks per hour, welders weld around one metre of plate per hour. More sophisticated constants can be derived, e.g., bricklayers lay 75 bricks per hour when building cavity walls above 3 metres but below 6 metres from the ground.

These constants allow for normal downtime – they show that a bricklayer lays 80 bricks per hour for which he is paid. Such output igures allow for teabreaks and other non-productive time and can be arrived at simply by observation. If you watch a bricklayer all day he will do all the things bricklayers normally do: make tea, climb scaffolding, leer at women in the street, adjust his clothing and lay the occasional brick.

You can therefore always allow for non-productive time in the production rates you use. The danger here is that those in higher management positions will display their prejudice and say things like 'You mean our bricklayers only lay 70 bricks per hour, I could do better with my hands tied behind my back.' Patient explanations about non-productive time being allowed for in the production constants will be like a  rubber skate – they will cut no ice.

*Add continuous 'background' tasks*

These tasks might perhaps absorb the first 10 hours of each resource's time for each week. To keep everyone happy you plan on everyone doing their standard week as paid for. They are available for, let's say, 40 hours per week in line with their terms of employment. You then introduce a high-priority, continuous task-absorbing 10 hours of every resource's time. This takes away the time spent on non-project-related work before you can begin allocating the rest of the time to tasks in the various projects.

*Plan specific downtimes as tasks*

This is neat and works well in some organizations. It deals with specific and unusual downtime periods rather then the continuous background non-productive or lost time. You create a 'phantom' project called 'training' and create tasks called things like 'attend programme-management training course'. This has a duration of three days and absorbs 100% of the time of the people going on it.

You can have another phantom project called 'holidays' which contains similar tasks, each of which absorbs 100% of the resource's time and is a very high-priority task. Descriptions might be 'Joe goes on leave' or 'Mary gets sunburnt again'.

Tasks in the 'training' and 'holiday' groups absorb resources just like any other tasks and, being of a high priority, leave nothing left over for the resources to contribute. The spin-off, as Murray Walker would say, is that you can produce barcharts from these two phantom plans showing everyone's holidays and absences on training courses as part of the planning service you thoughtfully provide.

## Implementation: the organization

This ain't going to be easy. You might think that all you need do is nip along to your local software store, buy a box or two, strip off the shrink wrapping (what is that stuff for?) and install the software on the selected computers. You'd be wrong. Sorry.

It is going to take some time to implement a programme-management software system simply because it must integrate a number of projects over some kind of transmittal system. If you have in mind to buy a single-user copy of Microsoft Project or something similar skip this next bit. I'm talking about a multi-user licence to use some appropriate software over a distributed network of computers and terminals. I'm talking about an integrated multi-project system.

Let's talk about the six stages of getting such a system up and running. At any stage you might think it appropriate to get up and run down to the Job Shop before the system comes crashing down on your head.

**Stage one:** the authority
A person or a group of people must have authority to get this system going. Generally a working party or committee gets set up to examine the issues and they either get the authority or report to someone who can sign the cheques. I am very much in favour of involving the people who will actually use the system so that they 'own' the solution whenever it is eventually selected. You cannot sensibly involve everyone but it is very easy for people who were not involved with the original policy and product-selection process to bitch and moan. Some people will bitch and moan whatever you do so there is no saving them.

For the same reason I am not keen on bringing in expensive and smartly-dressed management consultants. You might want them to answer some questions for you but they want to raise new questions so that you can employ them to not answer these new issues.

If you organize a working party you will soon have a body that represents the future users of the system gathering feedback about the needs and desires of the organization. Give this body backing. Everyone I have met who has been involved in installing a successful programme-management system or procedure has expressed the importance of their senior management's backing. I have included some case studies in the book and in many cases the interviewer speaks of the importance of senior management backing, of finding an enthusiast on the board.

Here is a letter. Show it to who you wish.

The Olde Cottage
Planning Approach
Critical Path
Projectsville
PV123 7FX

Dear Senior Director,

Please leave the choosing of a programme-management system to the people who will use it and give them your complete backing.

Yours faithfully,

Geoff Reiss

**Stage two:** the approach
The first job of the working group should be to evaluate possible approaches. Think about how your organization currently manages the programme and how it could be run better. Can you run some pilot schemes to test your ideas? Can you inspect a sister company, a subcontractor or supplier who has dealt with this problem? Have you got anyone who has recently joined from a programme-management company?

The questions here are not about tools, but about approach and strategy. Get this right and everything else will follow with, like the trapeze artist, the greatest of ease. Read whatever you can about the topic, go to the conferences and scan the magazines like *Project Manager Today*.

About the worst thing you can do is to dump the whole problem on your IT or IS or MIS department and get them to sort it all out. As this is not a system for the IT department, they are the worst possible people to deal with these questions. Unless you are an IT department, in which case please yourself.

I am sure there are IT departments who come up with very good solutions but the majority do not. Their horizons generally stop at computerizing what you are currently doing. It is their role to take the current process and add a layer of computer. This skips the approach stage and dives into the deep end long before we decided to go swimming in the first place. Possibly before anyone filled the pool.

So you think about approaches, strategies and how the system could, should and might work. Then select the best system for your organization. Then finally choose the software to support that system. These are meetings that you are organizing and perhaps running.

**Stage three:** buying it

Increasingly these days organizations are formalizing this process. Here you might call in a consultant to prepare a tender document which will specify the performance you expect from a programme-management software system.

Formal requests for quotations (RFQ) or invitations to tender (ITT) form roughly 30% of the project-management software marketplace and this is growing. The programme-management marketplace is not separated out from these figures but I suspect, if it was, formal quotations would play a much bigger role.

These RFQs list the functionality of the software required. They indicate the number of tasks to be handled in a number of projects, the number of resources, users, calendars and suitable response times and specific feature lists. They also list the training and consultancy services required as well as user-support arrangements. They rarely include hardware requirements.

Read what GEC Plessey went through to select a system in the case study. The case study is in the section about project-management tools in the programme-management environment. They approached it as an engineering problem.

Warning. Warning. Red Alert. Many organizations over-estimate their needs. This is what typically happens in such a company:

A survey form is done to evaluate the users' requirements. Great. A small group of people are given the task of catching everyone at their desks and asking them a few questions about project planning. Great again.

Fashion being what it is, most people exaggerate what they do. You find people who in reality only sketch out quick barcharts think that you will expect them to be doing loads more. It is a classic Emperor's New Clothes syndrome where everyone thinks they should be planning in much more detail and with greater sophistication than is really the case. And because everyone pretends, everyone pretends. Not so great.

I don't know why this is the case but many people try to pull the wool over your eyes with yarns about resource-smoothing algorithms, multiple calendars and complex interlinked subprojects.

The result of this sort of thinking is to over-specify your programme-management software. If you show a potential user a list of features that are available you'll get nearly all ticked and rated as important. The result of this is that a very heavyweight programme-management system is purchased and installed on the basis that it offers more of the desired features than anything else. It also offers extreme hardness of use, and puts everyone off getting the system going but it does offer all the functions that people said they wanted but didn't really.

You should try to keep it simple – the KIS philosophy.

The reliability of the software house is a significant factor when dealing with such a system. This is not a wordprocessor that you can dump tomorrow and replace, it is going to play a significant role in your daily business and you are going to need changes and help as your organization changes.

You'll go through the formal tendering process, receiving quotations, inspecting the software, running sample plans of your own through the system, meeting with existing users and comparing features. Eventually you will choose a system or combination of systems and give someone a happy hour with your order.

Somewhere around this time you have decided what kind of support team you plan to employ to look after the system. This can range from zero to a full-time team of five working all day on keeping files up to date and supporting users.

Generally at least one engineer or designer who is currently expecting to become a user will have become very enthusiastic about the system.  Because these people understand the needs of the users, are known to them and hopefully are respected by them, how about putting them in the support team? Maybe this enthusiastic person is wearing your clothes right now.

You might be able to significantly reduce costs by setting up an internal support team who will be the only ones to contact the supplier's user-support operation. Channelling user support can decimate the annual costs of looking after your many users.

**Stage four:** installing it
The supplier will do this as you have sensibly included the installation stage in your quotation. Unless you have (or are) a group of people who know all about this kind of thing, get professional help installing the system over your network or minicomputer system. It takes time and costs money but you should reasonably expect to be advised when the system is up and running.

The support team might be usefully employed to watch this process to learn about the way this system goes together.

**Stage five:** taking over the workload
Next it is time to stop messing about with sample files and get some real jobs onto the system. Generally organizations seem to take the view that, from day one, all new jobs will go onto the system, allowing current jobs to see out their lives under the old regime (if any) before dying of old age.

This is very sensible as the workload on the new system gradually grows as the new projects roll in. New projects follow the corporate line and methodology and are available to all through the new system. No one overloads the new system. Perhaps some long-term jobs might be added in to reduce the period of time when the two systems are running in parallel.

The disadvantage of this 'roll-on' approach is that benefits are hard to find until all projects are 'live'. Whilst some projects are not on the system the resource managers have to examine the workload of the live projects and add in manually the rest of the work. The live projects do not get any real benefit in terms of better-planned workforces. The senior management can dig down and examine any project on the system but have no such luxury available on the older projects. Where do you think their beady eyes will go first?

So in the interim period benefits are few and far between and the early projects are open to close scrutiny from above. This is a trying time for all. Throw in a few problems with the software or, worse, the hardware, and you can find people ready to give up and bin the whole thing just when the benefits are due to arrive. You need some special qualities to get through this stage.

**Stage six:** showing benefits
The wise programme-management person gets prepared to show the benefits of the new wonderful whizz-bang system the company has invested so heavily in.

The wise programme-management person starts working on benefits long before the system is installed. The first thing you need to do is to establish how well the organization is doing prior to the new system. Once it is installed and working you hope that there will be fewer problems, less panic and a smoother way of working. Does anyone realize that they have left the bumpy path and are now travelling on smooth ground? Hardly at all – they have far too many other problems to deal with. So you had better get an image of the state of the company before the programme-management system gets installed for later comparison. This is going to be your benchmark.

*Subjective analysis*

How about a survey of your colleague's views on the way projects go? You could design a simple form and make the survey vaguely scientific in an attempt to put some numbers to the feelings and impressions people have about the way projects are historically run. This can be camouflaged as a survey to help better understand the need for new methods and indeed can be combined with such a study. Here are some ideas for questions:

## Norfolk & Goode
## Project-Management Survey

1.   Interviewee's name
2.   Interviewee's position
3.   Brief description of function
4.   How many projects are you normally involved in as part of the project-management team?
5.   How many projects are you normally involved in as a resource?
6.   How well do projects generally do in terms of timescale?
7.   How well do projects generally do in terms of budget?
8.   How well do projects generally do in terms of meeting the need?
9.   Do you feel you know what each project aims to achieve?
10.   How well do projects generally succeed in meeting their objectives?
11.   How efficiently does the company use its resources?
12.   How well do you understand the priorities between projects?
13.   Are you satisfied that you can find out the current state of a project in terms of schedules, budgets and progress to date?
14.   What specific areas of project management would you like to see improved?
15.   Does any particular group have a problem with the project workload?
16.   Does the project workload impact the non-project workload?
17.   How much time do you spend searching and negotiating for resources?
18.   How much time do you spend on liaison with interdependent projects?
19.   How much time is lost due to not identifying interdependencies?

If you can run something on these lines by enough people in the organization you can draw some mathematical inferences from the data you collect. To get a really black position ask the same questions of the people who are not part of the programme-management team but happen to be the people who suffer when a project goes late. They should not be hard to find nor hard to get chatting. They'll probably love venting their feelings. Let them.

You could give each question on a questionnaire like this a scale of, say, one to seven, making it clear what the numbers refer to. You might decide that one equals very well, seven equals very badly and four indicates no strong view. Your questionnaire can be made up of statements rather than questions. A statement might say something like 'projects in this company generally finish on time'. One can equal total agreement with the statement, seven can indicate total disagreement with the statement and four shows no strong feelings.

This allows you to produce all sorts of fancy graphs and piecharts thereby showing a fairly analytical approach. If you think that you can get away with it you can later explain that you used the methodological orientation techniques propounded by Charles E. Osgood in *The Measurement Of Meaning* in 1957. This is the clever chap who came up with the scale of one to seven by simply applying the 'Rule of Six' – the number of things we can store in our minds.

You might also explain later that this form of questionnaire forms part of a structured interview which is classed as indirect observation. This grand-sounding bit of jargon means you ask someone what they think rather than making observations yourself. Structured interviews were regarded as very useful in testing a set of hypotheses by Sjoberg and Nett, who wrote about this kind of thing back in 1968.

By these means you can get a statistical analysis of the subjective ideas of what people think long before a programme-management system hoves into view over the horizon. You can expect people to be fairly damning about the whole thing and you should get a pretty black picture of the status quo.

| Status Quo had some black pictures on their albums. |
| --- |

Keep all this stuff. You might use it, or part of it, to bolster your case for an investment in programme management but keep the data for later.

## Objective analysis

Is there hard data you can get your hands on? Perhaps you can get access to old project files and build a picture of original targets (delivery dates, budgets) and actual achievements. This might show some very worrying trends and might give you a great deal of hard evidence to prove that things could be loads better.

Perhaps you can quantify the cost to the company of the inefficiencies and loss of business due to overdue projects and overspent budgets. How about cancelled projects? There is usually a secret cupboard somewhere storing tales of projects that got thrown in the bin to gather dust. If these were simply money down the drain you might be able to quantify the cost of cancelled projects over the years.

Take care with your career. You might upset some very important people who ran some very unsuccessful projects in the past. Perhaps it would be politically wise to avoid reference to specific projects and to stay with the mathematical analysis of many projects.

Quoting details of specific projects can be very dangerous. If you ran the job yourself you are running yourself down, if the Vice President – Projects ran that one years ago whilst she was a lowly project manager you will be unknowingly running her down. You are not even on safe ground tearing a past project to pieces whose project team have moved onto to other companies – there will be someone who still feels sore about it. Perhaps it is better to stick with impartial analysis of data. Equally it is better to deal with problems, difficulties and barriers than to deal with blame – you'll keep the collars cooler.

So, by a mixture of subjective and objective analysis, you have collected a catalogue of disaster. This will:

1. support your case for a programme management system;
2. come in really handy later.

When do you think this data will come in handy?

Some time later, once the system is up and running and most if not all work is on the system and when the attitude and morale seem reasonably high you simply repeat the process. Ask those questions again, examine the recent projects and do some even smarter graphs and three-dimensional piecharts showing how things have improved.

You can show the benefits as best you can, armed as you are with statistics and impressions. If they look good blow your trumpet. If they look weak, keep quiet and find out what is going wrong. You should be able to show that the savings created by the increased efficiency are greater than the cost of the system.

### It is a project, you know

The process of evaluating, selecting, installing, training and supporting a programme-management system is a project in itself. No one in the organization is more at risk for a little ribald jousting in the canteen than you if the project goes wrong.

Can't you hear them now: 'How's the project to get projects working, working?' Snigger, snigger. 'He couldn't organize a booze up in a brewery.' 'She couldn't organize an Italian election.' You had better make a good job of this project using all the right techniques, clearly stating and understanding what you are trying to achieve and planning to do so.

### Training

It is likely that you will run some series of training events as part of the move to programme management. Firstly there are two levels of training available:

- how to manage a programme;
- how to press the buttons.

The first may be a little contentious as most people will be of the opinion that they individually know exactly how to run a programme thank you very much, it is all those other layabouts who keep on screwing up. Such people will not wish to attend a basic course unless it is sold to them on a better basis.

It might be sold as a 'workshop facilitated by an expert in the field'. A workshop can mean that everyone contributes and gains so that the big-heads can think of the event as their turn to give whilst others learn. This may be far from the truth but it keeps some people happy.

The second is much less contentious and what everyone expects. They anticipate hands-on training with the software so that they can learn the right buttons to press, the mice to click and the meaning of those strange words computer programmers seem so keen on displaying in little dialogue boxes.

You might sneak some general education about programme management into a button-pressing training programme or integrate the two openly offering a course split into two parts. Generally, in training, people's fingers itch to get on the system and start producing results as soon as they see a computer. This is strange because for the rest of the time they can't get far enough away from one – perhaps a psychologist can explain.

You may find it sensible to structure courses for different levels of management. Generally firms have a series of courses starting with a long detailed one for those who will be users, passing through summary courses for senior management and lunch-time overviews for the board.

If you read the GEC case study you will read how very often people at the lower levels support the ideas behind programme management but feel that it will only work if their bosses support and understand what is going on. Many a boss just saunters around picking on people and giving them a hard time on the basis that this is the only way to ensure successful projects. This person will not be mollified by barcharts and histograms. This person will not even be mollified by success. Some organizations get an external training company in to run these courses, some run their own. There are benefits to both schemes.

External training companies are supposed to know the nature of the industry but cannot be expected to know much about the nature of your business. They are especially good at general courses. You might get a training company to spend some time with you within the company absorbing the way your group works and picking up your terminology.

Some training companies hire humans who will stand in front of a group of a dozen eager engineers and repeat verbatim a training course on project management with all the depth of understanding of a video cassette player. Such people even have a rewind button somewhere concealed on their body. Try asking a question, wait for the 'we'll get to that point later' response and then ask it again later.

External training companies are supposed to employ people expert at training. The ideal trainer knows the industry, knows the subject, knows your company and can teach. The worst person cannot teach and doesn't understand the industry.

Ideal people are going to be rare. Training companies have to mix professional trainers to whom they teach project management with project managers whom they teach to teach. You could do the same. You could take a few interested specialists who are known and respected within the company, who know the company and the industry and turn them into trainers. They can go on courses on presentation techniques and be backed up by a professional training company who might help with the original training material and the first few courses.

You might lose a few interested specialists who are known and respected within the company. Someone is going to have to do something about their previous workload and they sometimes get quite taken with the training world and take it up professionally. Trainers tend to be extrovert.

Some organizations run training courses on the company premises. Using the conference room is cheap but subject to interference and interruptions. Messages will arrive from secretaries, assistants and bosses which will take people away from your carefully-structured course. Running courses in a local training centre or conference facility makes the course feel important and different and separates the delegates from their normal day-to-day pressures.

Given a big enough budget you might take the course to a hotel some distance away, or perhaps central to your many factories around the country. You can soak people in the topic as they are truly separate from the office and home, from the filing and cutting the grass. Choose a nice hotel and you can soak people in the swimming pool. This is very expensive although good deals can be done with out-of-season hotels. Larger companies have their own training schools in converted stately homes in parkland where the resident training teams are very likely to want to row in on the act.

Two little points about training. Training companies use the airline charging technique. The more people on the course, the more you pay. This is great business for the training companies as the course cost for each extra person is only increased by the booklet, the freebie pen and a lunch or two. You can strike good deals for a long series of training workshops.

Not all training is achieved by this kind of 'face-to-face' training. There have been considerable developments in computer-based training using videos, CD-ROMs and interactive graphics. There are published sources of information and you might well create a programme-management manual for your company.

Some organizations issue certificates or diplomas to people 'passing' the training courses and workshops. Some very strong organizations 'licence' people to access the system, a little like using a car. You go and learn and only then get access to the system. Such companies are not trying to tempt people to use the system, they are insisting on a reasonable level of expertise before allowing people on.

The training industry has its own complete language including such gems as facilitator – the guy at the front; delegates – the students; and experiential learning – where the facilitator gets the delegates to play a game or take part in a simulation.

I'm personally very keen on what the training industry calls experiential learning for this simple reason. If you put someone up in front of a class of a dozen engineers to start babbling on about a topic the delegates' minds will begin to wander. It is inevitable. Especially at two p.m. after a pint, roast beef and pudding, in those comfortable chairs with the heating on and the lights down low as the slide projector gently hums. Eyelids will droop as those minds wander off to the land of nod.

This is not an accusation nor an indictment, just a fact of life. Students often drift off to the land of nod during long lectures. People doing something active stay awake and alert much longer. Getting the students to actually do some work as a team with the guidance of an expert and perhaps after a short theory session is a very good way to learn.

To back the programme-management drive, some organizations get the chairman or chief executive to open the training courses with a short presentation about how important this all is, how much effort has gone into it and what the delegates can expect to get out of it.

User support

'Users' means the many people who spend part of their day tapping away at your wonderful, all-singing-and-dancing programme-management system. You will need to set up a system for supporting users of the system who get into trouble. There needs to be some internal mechanism available to those who use the system to get some help when they get stuck.

Many organizations have a team set up to support the programme-management system. They do not support the programme at all, just the system. If Fred cannot get the report he wants, if Liz wants to set up a report run which she can easily run off on Tuesday mornings, if the MD wants an A4 summary before every board meeting, they should know who to turn to.

The team might produce a little newsletter telling everyone who cares to read it about a nice work-around for an annoying bugette that Irene has worked out and who is using the system cleverly, what advanced training workshops are planned and what goodies are coming in the next release.

Such a team takes the brunt of the load but even they may need to go back to the software house from time to time. They may go back to sort out some new problem or bug, there may be a new release and they might fancy a job with the software house someday. Having such a team should keep your user-support costs down as very few calls go back to the vendor.

Such a team may have regular jobs to perform on the system like purging old files, transferring project-costing data across to the accounts people, like monitoring which timesheets have been returned and chasing those that are late.

There is more about choosing and using software in the programme-management arena in the section entitled 'project-management tools in the programme management environment'.

## Case study: Ford of Europe

I am indebted to Ford of Europe and *Project Manager Today* for permission to reprint this article. It explores the change to a programme-management culture that took and is taking place within Ford.

## On the road with project management

A modern motor car is probably the most expensive and complex piece of machinery you will ever control. Today's car contains so many diverse elements brought together into one reliable and easy-to-operate machine that we forget what an engineering achievement this is.

Reading that first paragraph might make you think of those vehicle-assembly production lines where both human and robotic workers fix bits to cars passing slowly by their work stations. What I really want you to do is to think back to the stage before that when a new car is designed, prototyped and tested.

Within Ford of Europe new vehicles start life as a concept, an idea in the designers' minds. There follows a lengthy process of giving that concept shape and form, of converting it into a manufacturable vehicle. Work is done by specialist engineers working in diverse departments in a number of locations in Britain and Germany. This process is called 'Concept to Customer' and involves all the complex stages between the birth of a concept and the first contented customer driving away in his new car.

It is in this area that Ford have invested considerable time and effort in an attempt to get products to the market more quickly – you can readily see how important this is. Project management tools and techniques have been employed within Ford of Europe to plan and monitor the whole series of development work and this is that tale.

Hence our trip to Ford to find out more about this development process. We travelled to Basildon to meet Ford projects people Nigel Booth and Alan Banks who led us willingly through their world of vehicle development. They are very cautious people and are careful when talking about the development process and the future cars, in fact anything that may be useful to their competitors.

By the way, do you know which is the most popular car ever built? Answer at the end of this article.

### Project life cycle

Traditionally Ford has been made up of four major functions: Engineering, Sales & Marketing, Manufacturing and Finance and these functions lived and worked within their functions. Ford have been trying to throw out the walls between the departments in an effort to improve their ability to develop new cars quickly and effectively.

Nigel Booth – Manager of Pre-production Systems in Basildon and Alan Banks – Co-ordinator – have been working hard to introduce a Total Programme Work Plan as a part of Ford's 'Concept to Customer' programme. TPWP is a methodology which is supported by PSDI's Qwiknet software and, combined with training, consultancy and some management restructuring, is designed to project manage the whole design process literally from Concept to Customer.

The need was to shorten this development cycle and improve quality partly because of the complexities arriving as world-cars and European cars become the norm and partly because Ford's Japanese competitors have the annoying habit of getting their cars out on the market with uncanny speed.

The development process is broken down into phases separated by milestones. We are here dealing with the main programme from the moment when the commitment to build the new vehicle is made through to another milestone when the first customer drives away from the showroom in his shiny new acquisition.

Each phase starts with a set of information and ends with a set of deliverables. At the end of each phase is a 'gateway' through which the project can only proceed with the relevant approvals. Each gateway is marked by a milestone like the two mentioned above. Readers from the IT industry might recognize this as a methodology where the names and terms have been changed to protect the innocent. It does seem that the concepts of a methodology – a structured and organized method of working – are spreading out from their Information Technology roots.

I do believe that a methodology has as many dangers as benefits. The benefits are that a set of guidelines are established that are aimed to bring some order to the project life cycle. Everyone understands what stage the projects are at, what information is needed to start a phase and what deliverables are expected at the end of the phase. One danger is that the methodology is inappropriate. It is not uncommon for a team of enthusiastic but green graduates or a team of near-retirement projects people to be given the task of designing the project life cycle. The methodology is then rigidly followed by the project teams because it is the approved way of doing things.

Another danger is that the methodology becomes inappropriate as time passes, new techniques are developed and the company structure changes.

Methodologies need to be followed but they also need to be reviewed from time to time.

Once the Ford development process was designed the next stage was to develop a system so that this process could be speeded up by use of simultaneous engineering. In the old days each process was carried out before the next began and simultaneous engineering refers to a co-ordinated overlap arrangement between the various functions. Then things started to get really complicated.

### Work-breakdown structures

Ford is packed solid with expertise and their work-breakdown structures (WBS) reflect the depth of work that goes into each new car. The WBS has a new car at the highest level and this might be broken down into engineering activities like chassis, body, electrical systems, power train and so on.

Additionally the vehicle is broken down into component systems – an example would be a head lamp. To manufacture and fit this component into the vehicle requires input from at least two engineering areas – body and electrical. Most components interact with two or three engineering areas.

There are people in Ford who devote significant parts of their working lives to dealing with items such as door locks and they are lost deep within the engineering structure of the organization. At any one moment a door lock engineer might be involved in an Escort Replacement development project, the concept of a new Granada and the update of the Fiesta.

So we have here a jungle of functions and components linked by engineering functions trying to develop a series of complex products. You can see that project management might have a role to play. At Ford each new vehicle is referred to as a programme rather than a project.

Team Leaders play a key role in getting work done within Ford. Team Leaders are almost equivalent to project managers in that they have no resources of their own but get their work done through Supervisors who have specialist resources working on a number of programmes. The Supervisors assign engineers to each team leader for each programme.

This means that each engineer may have two bosses: an engineering manager who deals with career and personal matters and a team leader who deals with the engineer's output.

The ideal solution to this might seem to be the creation of a single team to work full-time on each programme together with all the other engineers. This way you get a single-minded team co-ordinated and working together to a shared objective but there are problems with this concept which is known in the motor trade as the Co-location approach. Firstly there must be one team per programme, which is very expensive. The team members' career path is a problem as is the need for like-minded engineers to get together and chew the fat. How do you get all the door lock specialists to develop new concepts in door lock technology when they are busy working on their own programmes? How do they grow within their specialization and become Senior Door Lock Engineers?

Another problem with Co-location rears its head as the project nears its end. The whole team will need a new project to move onto when their current programme ends. This is a problem faced by major builders and engineering companies worldwide. I guess that if there was a nice elegant solution we wouldn't be talking about the problem.

Nigel and Alan's role is to support the engineering programmes. They don't actually do much planning or hands-on project management but support the people that do plan and manage the workload. They are an internal project-management consultancy.

'We break our work down into a number of areas', explains Nigel. 'First we have an education process to extol the benefits of work planning throughout the organization. We found some planning on paper being carried out but the results varied from team to team. We felt a common approach was needed.'

So Nigel's team set out designing an internal project management software system aimed at meeting the needs of the leaders and supervisors – the programme side of the management matrix.

They set up a few pilot schemes which went well so now the whole programme is being rolled out. The budget for project management-software, plus computers to run it on, plus consultancy to meet their needs on one current vehicle programme is $2 million. It is big business this car business!

The institution of this system was a project with a project plan and budget and I know many people that would think of such a project as a very big one indeed. Alan monitored and controlled the project like any good project manager should. Nigel and Alan organized the installation of about 120 PCs for about 200 to 300 people and this equates to about 50% of their estimate of Ford of Europe's total need in terms of engineering support.

Ford in Europe have followed their American cousin's lead and committed to PSDI's Qwiknet Professional on the PC. They maximize their use of the software by consolidating the plans at a high level using Qwiknet's multi-project facilities to examine resource overloads and to monitor progress.

For example a detailed plan for a door would be consolidated into its 'homebase plan' – in this case body engineering – and later into one large plan for the whole programme.

Each element on the plan – e.g. a head lamp – has its own PERT chart with links to many other plans. These links become significant when the many plans are merged into one so that the interfaces can be checked for conflict.

Another benefit derived from the overall plan is the ability to produce a master summary of the plan, cross-referencing various engineering groups (body, power-train, electrical) in a report to the senior management.

This sounds a little like total automation of everything in an all-singing system. 'Not true', says Alan,' we don't automate everything but we do encourage people to talk to each other, to ask questions. We sell the system as a model with which we will tear down the chimneys.' If you, like I, thought that chimneys were parts of a house and not a car you should understand that the concept relates to engineers working in isolation, separated from their colleagues and other interest groups and only meeting high above the basic project in hot air.

### Potential Fail Zone

One early problem explained by Nigel was the Potential Fail Zone that occurs early in the life cycle of a project.

If you take any project and draw a graph of the effort you expect to put into project management against time, you would get something like this:

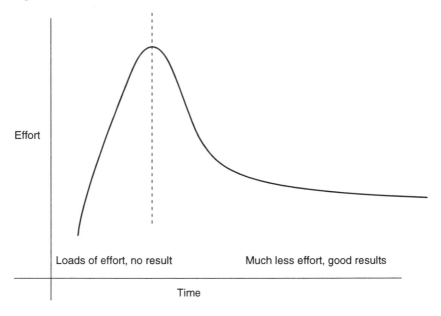

The graph shows that a great deal of effort goes in early on in the project life cycle as the original planning work gets prepared.

The highest point on this graph is the Potential Fail Zone. You have invested generous amounts of money and time and there are few if any results to show. You have just finished, if you excuse the Biblical analogy, tilling the soil and spreading the seed. Any minute now you will start to see the result and reap the harvest as you move from planning increasingly into monitoring and control. This is when your senior management, frustrated at the lack of apparent results, are most likely to pull the plug on the whole damn project management business.

Ford get help on the way up to the Potential Fail Zone – they buy in knowledge in the form of consultancy to help them get over the hump.

This kind of implementation is not likely to be an easy process and we asked how training had been handled. Nigel responded.

'There were two schools', he replied, 'we train people to understand their role and position in project terms and sell the idea of giving and getting. We also teach the techniques like PERT, WBS and so on.'

What does Booth mean by giving and getting? He explains: 'many people in projects have to do extra work to provide data and get little or nothing back in return except complaints about delays. Ford try to explain how everyone should give data and everyone should get a better idea of the plan.

'It is a missionary job', Nigel continues, 'and the workforce is becoming more informed. We are training senior managers in work planning in a two-day course and have a package of information for very senior managers to read and absorb.'

After a major investment which any project management software, training and consultancy company would like to get its hand on, is it all working?

'It is really a little too early to say, come back and ask me in a year or two', smiles Booth.

We will wait until the system has been up and running for some time, until a time when a couple of new vehicles have been launched quickly and effectively. Then Nigel and Alan might be able to tell us more about their successes.

Project Management appears to be bringing about improvements within Ford which is, after all, a classic project-management environment. Ford does offer good examples of both ends of the spectrum – project and production management. Whilst Nigel and Alan help in the project-management arena there are no doubt others trying hard to improve production techniques.

It is rather unfair to ask Nigel and Alan if the techniques are working as they might result in speedier car development processes, better cars, lower development costs or a mixture of all three and other factors are simultaneously and continuously improving Ford's way of working. One thing you can be sure of –

Nigel and Alan are doing their bit to bring project management into a production environment.

The world's most popular car:
You might think that the mini, the Morris Minor or the VW Beetle were pretty popular but the most popular car ever built was the Model T Ford which made Henry Ford's fortune. It was available 'in any colour as long as it was black' and has yet to be outnumbered by any vehicle. It went on sale on Oct 1, 1908 at $850 and a stunning 15 000 000 had been made when production stopped in 1928.

## *Budgets, cost control and things financial*

### *Money, I hate the stuff*

I like the things I can buy with money but it is useless in itself. Sure, you can stuff a mattress or wallpaper a wall with old Russian million-rouble notes but you can't eat money. Dosh is an illusion – it is really only the bits of paper you use as part of a complex international barter system. Money represents the debt society owes to you for making your contribution.

If money in your pocket is an illusion, money inside a computer is trebly so. It is not so important, it just happens to be the score by which we measure success and that has been made important. Now the lack of money – that is a problem.

But many people regard money as the central issue of programme management so we had better take a look at the various ways in which you can get into trouble and ruin your promising career by running hugely successful projects which look black on the balance sheet.

> That's black as in black look, not black as in bank statement.

### *Project approval*

If you are involved in running a programme of projects which are aimed at achieving the goals and objectives of the organization you are following the CCTA definition of programme management. Organizations that find the CCTA's work appropriate have names like North East Water, Huddersfield Underground Railway or The Uncivil Service. They are normally paid for by you and me through an ingenious system called taxes.

In such environments the projects are not aimed at creating a deliverable which will be delivered to a client and paid for, they are aimed at achieving the long-term goals of the organization. Water companies build new sewerage plants, railway companies lay new track and quangos quang. These projects are designed to make the organization better, more efficient, improve morale, create a new asset or give a better service.

In these environments the first step for the budding project manager is to convince the powers that be that they should invest huge sums in this new and innovative project. Actually that is the second step, as the first step is dreaming up the project that is to inspire such enthusiasm.

The aspiring project manager person puts together a case promoting the value of the ingenious project, showing how the investment will result in a golden future for the organization. It is quite normal for a company to allow individuals to do preparatory work on a project – enough to get together a business case for the project.

In some organizations there is a definite system to encourage people to come up with ideas for new projects and to permit the time to develop the ideas into a business case. Pilkingtons, the well-known see-through manufacturer, actively promoted free thinking and innovative projects by a business-development programme. Engineers were encouraged to dream up projects that would develop the business. Some projects were designed to lead to new products but many were more to do with efficiency and economy.

Armed with the business case the project manager then has to try and get approval to proceed with the project and this means approval to spend the necessary money. Make no mistake, it is the money you are really after. Most managers will happily approve your project if you offer to pay the bills yourself, but will be a little more dubious if company funds have to be set aside. This process will put you in front of some very influential people.

There are a number of problems that will appear in this approval process and these problems will manifest themselves as spoken words. These words will be spoken by a senior manager with a forked tongue. Some of the reasons you will be given to prevent your project from proceeding are mentioned below in order of increasing truthfulness:

The company may not have enough money
The predictions for income over expenditure for the forthcoming period do not leave enough cash to pay for the super project you have dreamed up. This excuse does not suggest that your project proposal is in any way faulty, only that the expenditure cannot be borne. Will the company have enough to pay your wages?

The company cannot afford to borrow that much
It is a regrettable fact that if you go cap in hand along to your bank manager and lick his shoes, compliment him on the nauseating picture of him, his wife and kids and remind him that this is the bank that likes to

say 'We're listening', he (and it is usually a he) may lend you some money.

He will only lend you money if he is of the opinion that you don't need it. The more desperate you are, the less chance you have. However much you borrow they do expect you to pay it back at so much a month. Sometimes your company cannot pay back enough to support a suitable loan or cannot raise the capital in the first place. If this news reaches your ears, it may be a good time to move to work for a company with a better credit rating.

### Your project may not show a high enough return

If the financial people insist that money is to be the target you should be able to show how investment in your project will cause an upturn in your company's fortunes. This might be in the form of reduced outgoings due to better quality control, less wastage or perhaps a more efficient manufacturing process. The upturn might be in the form of additional income generated by the capital asset which is the goal of your project. At least we are getting nearer the true reasons for not giving a project the go-ahead.

### Your project might be too risky

As a part of the case you put forward to substantiate the essential rightness of the proposed project you should have identified some risks. The head of the Accrington office has a future that would look bleak under your proposals to merge the Accrington office staff with the local job centre applicants. He has also identified an impressive list of risks each of which is unlikely to occur or of infinitesimal impact or both. You know this, he knows this but the board members do not.

The board reads through this list of risks and gets the collywobbles and are now giving the project the cold shoulder and giving you the honourable and long-respected 'being discussed' delay. The truth here is that no one wants to speak up on your behalf in case those horrid risks come to pass.

### Your project is not a high enough priority

Now we are getting near to the truth. There is a reasonably fixed sum of money that is available for investment and your project is not viewed as being the best. All those projects ahead of you in the queue have got (or appear to have got) a better return on investment or some other attraction that makes it worth while for the board to go for them rather than yours.

The other attractions include important elements like: if we have a factory in Spain I can go and visit it and take a few days off in the sun; if we had a wholesale outlet in Chelsea I can go there on Saturday mornings before the match and claim my expenses.

We just don't feel comfortable with your project
This is generally the real truth. If they felt comfortable enough with the project they would find the money, borrowing if necessary. If they liked your plan they would move your project up the queue, take the risks and go for it.

All those cover stories about insufficient funds assume  that there is a ceiling, they know where the ceiling is and the possible projects fighting for the money are in a sensible order.

It is like you saying that you cannot afford a new bedroom carpet. In English this means that you are not willing to move the bedroom carpet further up the list of items currently ahead of it in the queue for money which probably includes: food, ale, cosmetics, expensive bicycles, holidays, alimony and more ale.

You have to sell your project if you want it to succeed. They have to believe in you and your idea. The chances are they do not have a fixed ceiling on capital investment but a moveable target which can be moulded to suit the prevailing conditions. It is also very likely that they are not very sure how much is currently spoken for in forward investment.

If the board wanted to know what jobs were being planned, approved and monitored they would prepare a set of figures that look something like the diagram below for the company's investment strategy.

The idea here is that a mix of projects are all vying for life in a highly-political survival-of-the-fittest competition and every few weeks some projects get cancelled, some new one come along and existing projects change their budgets and timescales.

The table on page **122** lists the projects down the left-hand side stating the project's code number and the names of the project manager and senior project manager. Alongside that is a description of the project sufficient to identify it and a status marker. In this company a project's status can be any one of the following:

**Proposal:**  aproved to spend internal resources on project evaluation only.

**Design:**  aproved to proceed through design stage using internal and external resources.

**Committed:**  aproved for all stages.

**Completed:**  investment stage ended, returns expected.

Your employer might have different status levels. Your employer may not have thought of status levels at all in which case you might get noticed by passing similar ideas into the management maelstrom. Many companies insist on a formal process to permit movement from one stage to the next and this might involve a presentation to the senior management at which viability is open to probing questioning. It might alternatively involve a presentation to senior management at which prawn sandwiches are open to questioning probing.

## Norfolk & Goode three-year capital programme

| Project code | Project manager | Senior project manager | Project title | Project status | 1995 | | | | 1996 | | | | 1997 | | | | Totals |
|---|---|---|---|---|---|---|---|---|---|---|---|---|---|---|---|---|---|
| | | | | | Q1 | Q2 | Q3 | Q4 | Q1 | Q2 | Q3 | Q4 | Q1 | Q2 | Q3 | Q4 | |
| 01/0976 | Fred | Joe | CAD system | Proposal | | 10 | 10 | 50 | | | | | | | | | 70 |
| 01/0978 | Sue | Joe | Flow-line production facility | Design | 30 | 30 | 234 | 456 | 375 | 300 | 250 | 200 | 50 | | | | 1925 |
| 01/0877 | Alan | Mary | New Northern warehouse | Committed | 556 | 456 | 654 | 300 | 245 | 100 | 45 | 20 | | | | | 2376 |
| 01/0866 | Julie | Keith | Own lorry fleet | Committed | | 50 | 50 | 50 | 50 | 50 | 50 | 50 | 50 | 50 | | | 450 |
| 02/1876 | Pauline | Dave | New staff canteen and dance hall | Proposal | | | 20 | 20 | 50 | 150 | 150 | 150 | 300 | 50 | 25 | | 915 |
| 008/456 | Dave | Keith | Replacement accounting system | Proposal | | | | | 25 | 25 | 100 | 100 | 100 | 50 | 50 | 25 | 475 |
| | | | | | | | | | | | | | | | | | 0 |
| | | | | | | | | | | | | | | | | | 0 |
| | | | | Totals | 586 | 546 | 968 | 876 | 745 | 625 | 595 | 520 | 500 | 150 | 75 | 25 | |

Figures are in thousands of pounds.

We'll come back to this question of status in a little while. Back to the table above. Alongside these projects is a matrix with a timescale along the top. For each time period, for each project, a sum of money is shown which represents the amount that is planned to be invested on that project in that time period.

Keith is masterminding the replacement accounting system project and this is so far only at the proposal stage. He expects to spend £25 000 in each of the first and second quarters of 1996, which is probably the design phase of this project, which is all he has approval for at the moment. After that, assuming the project proceeds, he expects to spend £100 000 per quarter for the rest of the year as the software is written and the hardware is installed.

These budget figures are not based on a detailed project plan with cash-flow envelopes and all that good stuff. They are very rough and ready, broad-brush estimates of the project's finances. The people involved know very little about the project in detail at this early stage, which makes detailed planning even less useful than normal.

Organizations that use tables like these tend to have a once-a-year round of project justifications. Once a year the projects people and the financial people get together to amend a table much like the one above.

## Project budgets versus annual budgets

This sort of table (the one above) deals to some extent with the old problems of project budgets versus annual budgets. This is similar to the old problem of people with vision versus accountants.

Accountants tend on the whole not to be great visionaries. They have a driving need to make things balance and would be great in the circus ring with a bowl of flowers, a couple of swords and a unicycle. Forward-thinking and forward-looking accountants do exist but you are unlikely to have one in your firm as they are all senior partners in the major accountancy firms.

The normal accountants' ideal is to keep things on a level keel and (to stick with the nautical analogy) know exactly how high the water is. They don't seem to care if the boat is sinking or not as long as they know exactly how much water is getting in.

A basic precept of accountancy is something called the tax year. This is a period of time equal to the period of time it takes the earth to go round the sun (which is convenient) but which starts and ends on otherwise unimportant dates. Only in the religion of the Sacred Heart of the Immaculate Double Entry does New Year come on the 5th April. Your firm has probably got yet another New Year's Day which was chosen by someone years ago in a drunken stupor who called it a 'tax year'.

This is all very well and of little significance to us projects people until you have a project that crosses the company's tax-year boundary. As far as the accountancy goes you have to predict and record exactly what you intended to and did spend before and after that magic date. You also have to make these long-term predictions about expenditure in each tax year and stick to them.

Finally, if the company system makes no special arrangements, your project is effectively up for review at every tax-year end. Your project knows little of tax years but deals in overall investment and returns. Your accountant knows of investment and returns but wants the books to balance each year.

Projects have been cancelled because there was a cash shortage for the coming year and the project got dropped from the budget. The project might have been 90% complete and only a few more pounds were needed in accordance with the original budget to see the returns flowing in, but the job got scrapped. You might have some sympathy for the project manager whose project, due to severe problems, over-runs into the next financial year. Crossing this dateline means little to the project team but there may be no allowance for this project expenditure in the following year.

So a system must be created within the company that recognizes both needs: long-term project budgets across year ends and the year-by-year financial forecasting and reporting beloved by accountants.

Even if such a system exists, if you have to spend the right amount each year to fit within an annual budget what happens to the project if you underspend or overspend in a given year? This can also create a serious conflict in inflexible organizations. You see the chance to get ahead of schedule and save some money by buying in some plant early but are stopped as that expenditure is not due until next period.

By the way, do include all the costs – many are hidden and don't count or are not counted within the organization. Some organizations work partly or entirely with person-hours and then multiply this by some figure representing person-time costs. These can be hugely over-estimated or, occasionally, way under.

It is worthwhile having a clear idea of who costs what. You can get into a situation where the project manager finds it cheaper to go outside to a contractor and pay real money out of the company's bank account than use internal resources. This is despite the fact that internally no real money changes hands but is due to the ludicrous internal cross-charging system.

In such a situation the project manager looks better if he costs the company real money whilst equally real money is paid to people sitting around with nothing to do. The only real problem was that the inter-departmental charging system is screwed up and makes the projects look really expensive.

## *Scheme approval by stages*

Many organizations suggest that all projects pass through a series of stages or phases and that project managers might stick with this phase definition. They suggest this in the same way as they suggest it might be a good idea to arrive on time each morning and not to take home valuable hardware. Some organizations relate permissions or approvals to these stages. Here is a typical stage approach from at least one water company:

> The words stage and phase are fairly interchangeable except in the theatre and electrical systems.

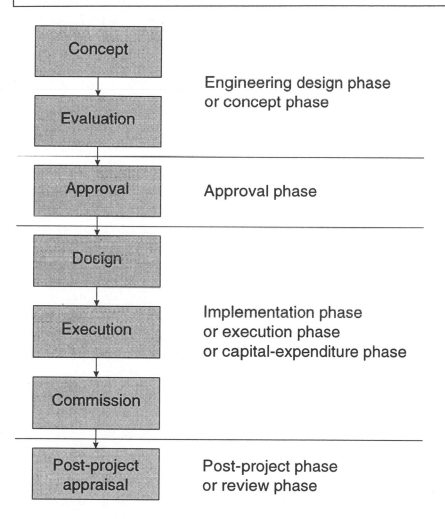

*Typical stages of a scheme*

In this company you seek and get approval to proceed through the first two stages which lead into another approval phase. Once approval is gained it is plain sailing through the design and execution phases. Some companies might prefer to approve the project after design but before execution. Some might insist on both.

If your company does not have a set phase process or does not have a graphical diagram of the existing process, you can take this diagram, muck about with it to suit your company and use it to slightly progress your career. Comments on how it went, written on used fivers, should be sent to the publishers for the author's attention.

We can get yet more sophisticated by adding in some elements of accuracy and risk and permission to these simple stages. Take a look at this example, which compares accuracy of cost estimate with project phase:

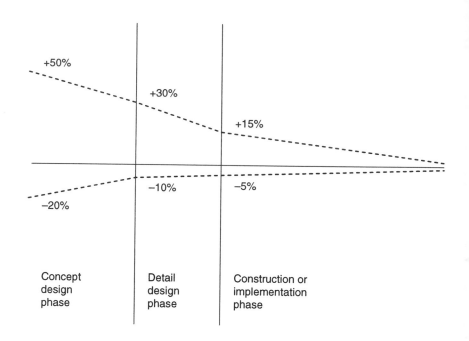

*Accuracy of estimate versus phase*

These ideas about phases are based loosely on BP Exploration's approach. In their environment the projects are large engineering thingies which look and smell so awful they have to be left out in the middle of the North Sea. The concepts behind this approach apply equally as long as there is a relatively cheap design process before an engineering or execution stage during which the real money gets spent.

In such environments new projects are divided into three main phases with names like: concept (or initial) design, detail (or final) design and the construction (or implementation) phase. The project team are likely to seek approval to move into each and every phase and indeed cannot move forward without that permission.

Initially a very rough outline is produced with some rule-of-thumb cash estimates. This is used to explain the nature of the project, the potential risks and return and to seek approval to move into the concept-design stage. Here a relatively small sum of money is requested, authorized and then spent with a group of internal designers hacking the thing about to get it into some sort of shape. At the beginning of this stage estimates are expected to be between +50% and –20% of final costs.

Once the internal design team have spent some time examining the project, a number of problems have been addressed and dealt with and a clearer idea of the content of the project has emerged. Armed with this information, the project manager revises and extends his financial justification for the project and submits the results of the work done in the concept design stage and requests approval to move forward into the detailed design stage. The submission would give a better idea of the estimated financial future of the project and would also include an estimate of cost for the next phase. By now estimates are tighter and expected to be within the range +30% and –10%. You might note how the project costs are expected to rise more than fall.

Once this approval has been granted the project goes out for detailed design. This would be a great deal more expensive – it might involve external design consultants, tests, prototypes and engineering experiments. At the end of this detailed design process a much clearer idea of the nature of the project has emerged and the estimate of cost has been once again refined.

Armed once again with this detailed design and a much more detailed cost and financial prediction, the project manager goes back to the powers that be for a final time to seek approval to actually construct the thing. The cost of the two design phases might tot up to 10% of the total project costs, so this last stage is likely to be requesting an amount of money nine times that previously committed.

The board are going to be much more thorough with their investigation and expect to be fed with more detailed estimates before approving. This final stage might require approval by a higher authority. By now costs are supposed to be tightened down a great deal within the range shown – a reasonable contingency at this stage.

You can see that any company can set up a project-approval process with a number of appropriate stages. At each stage the project manager seeks approval to proceed to the next stage and justifies that application with information gained in the previous stage. At each stage the degree of accuracy of estimates is expected to improve as the uncertainty reduces and often the cost increases. You might be able to search through the company archives and find out how accurate these kinds of estimates have historically been over the last few years which might well be hysterical historical data. At each stage the approval mechanism might involve more senior people within the organization. The further along you get, the higher up you get.

There is no reason to fix the system regardless of the size of projects. You can have a two-tier approval system, one for projects under £100 000 and another for those over the limit. At any stage, a project can be cancelled or shelved due to factors within the project and due to changes in the corporate environment. You can always take your project down the road to the opposition. Organizations that have a formalized procedure tend to allow people to come forward with projects at any old time of the year. Approvals might come from a meeting of senior managers and they may only meet every three months but there is no fixed annual project-review season – it goes on all year round.

Applied Business Technology International have a slightly different view of these ideas and I am grateful for David Marsh of that august organization for his sense of humour, common sense and also for the following idea. This shows how uncertainty over a project's future starts off being very large and drops away, finally reaching zero only when the project is complete.

Uncertainty in these terms could refer to the overall project costs, overall project duration or your chances of surviving with your job intact.

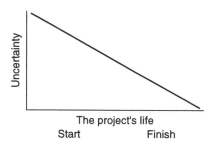

The project's life
Start                    Finish

## Survival of the fittest

I hope you see that in these environments it is a management jungle and the rule of the jungle applies: only the fittest survive. There are many projects out there all seeking to gain access to the precious money which will give them life. There is almost certainly not enough dosh to slosh around so only the fittest projects will go forward to grow strong and healthy.

Is it surprising that people are optimistic at this stage? Imagine, there you are working up a fairly detailed look into the future of your hypothetical project and you are thinking about how much it will cost, what the returns will be and what the risks are.

Are you going to be pessimistic, tell the bald untarnished truth and watch your project become stillborn or are you going to do what everyone else is doing, perhaps subconsciously, which is underplay the investment and risks and over-estimate the returns. It is as natural as being human and if you are not human what are you doing on this planet without a permit?

This is why many projects almost inevitably come out over budget and behind schedule. It's because people get enthusiastic about their exciting dream projects and do not bother to look under stones until they have to move a stone out of the way and find it is really the top of a mountain. For this reason some organizations have a system for reviewing the enthusiastic project manager's proposal. This can be a technical review group who go over every proposal analytically and dispassionately and add their views to your submission before it goes in front of the board. There can be a peer review where your fellow project managers take your project to pieces and give you a fairly hard time, again before the formal approval session. They can't get at you too much because you should soon have a turn to get a look at their pathetic project.

The word 'fit' in the phrase 'survival of the fittest', when applied to projects fighting for funds, should mean those projects which will have the best effect on the organization, those projects that will move the organization closest to its objectives. Actually 'fit' usually means those projects that the MD likes, that have a high profile or that are run by popular and well-liked project managers.

Some organizations have separate groups to evaluate potential projects. There might be programme managers, capital-planning managers or project-appraisal people whose function is to take a number of 'expressions of need' which could come from anyone in the company and translate those needs into a series of planned and integrated projects. 'Expressions of need' state a requirement – 'We really could do with 25% more capacity on the South East Region railways.'

The need is translated into a number of possible solutions (run more trains, run longer trains, two-storey trains, buy buses, shoot some passengers) then whittled down to some which become projects, perhaps one is to acquire the needed rolling stock. This process is more dispassionate and analytical but requires some considerable expenditure with full-time people busy evaluating projects. The evaluation group is soon likely to settle down into a rut without any injection of new blood and bright, agile minds in the rest of the organization don't get a look in.

Very large organizations have a hierarchical approval system where programmes are first approved in very broad terms and then, within those programmes, specific projects are approved. Alternatively, individual projects may have to pass a local approval system before going up to the regional headquarters and then finally the international HQ.

## Accuracy of cost estimates

It is sometimes wise to get the most accurate estimate you can. It is particularly wise if you are liable to be taken to task for poor estimating and those inhabitants of the carpeted offices do tend to come down a little heavily on the poor person who estimated a loss-making project.

I'm not sure if this is going to be a section full of useful excuses for things that have gone wrong, balls of wool to pull over people's eyes or some techniques that might help you get more things right. Let's see.

Here is a more general statement of the problems of accurately estimating project costs which is a lame excuse and should only be used in career-threatening emergencies: one of the fundamental problems of cost estimating is that the stage when a cost estimate is most useful is the stage when the cost estimate is least accurate.

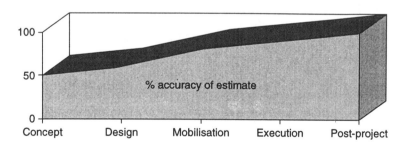

As you learn more about the project, as costs get firmed up by quotations from suppliers, as the actual deliverable becomes clearer and better defined, the accuracy of estimates increases. Many organizations state the degree of accuracy associated with any estimate and this is what BP are shown to do in the example a few pages back. This is a problem which everyone will recognize once it has been pointed out to them. However, if you use this as an excuse be prepared for the response 'Project managers with a future with this firm seem to manage.' This is a threat.

*Accuracy versus type of project*

This accuracy problem is worsened in strange or unusual projects. Where an organization tends to carry out certain types of projects it is reasonable to expect them to build a database of actual costs and be able to use that information to estimate future projects. You might reasonably expect that Barretts have got the costing of a three-bedroom detached house down to a fine art. It certainly looks like it anywhere but in the show house.

'Strangers' – a term used to indicate projects about which the company has very little experience – are clearly more difficult to estimate accurately. Barretts could do little better than you in estimating the cost of a bridge. Also, the more complex a project, the greater the difficulties in accurate estimating.

There are some ways of improving accuracy. Some techniques have been used to give first-order estimates for projects where some historical information is available. 'First order' might be the estimate you use to seek permission to proceed through the first phase or it might be the only estimate you ever do.

---

'First order' might be the first words you speak in a pub.

---

*Metrics*

This is a clever idea. The organization searches for simple measures of a project that relate directly to project costs. For example, in the shopfitting trade certain retailers use a very common standardized design so that you will feel comfortable and at home in one of their shops whichever town you happen to be in. You know what to expect when you open the door to Boots, M&S and a whole load of high-street shops. In fact, you know what to expect when you turn onto the high street, never mind going into one of those boringly similar shops.

These retailers can accurately estimate the costs of fitting out a store on the basis of three figures: retail floor area, non-retail floor area and length of shop front. They apply these three figures to a formula and, hey presto, there is a first-order costing for the shopfitting project. The factors you decide to use may be a little off the wall but statistical analysis will usually show the common relationships and will lead to quick and cheap estimates. You can only expect to apply metrics to repetitive projects (runners and maybe repeaters) and only then if you have built up a database of past projects to base your metrics on.

Some software people use metrics extensively and talk about the number of files, the number of transactions and the number of screens that need to be created. From such data quick estimates of the number of programmer days and a cost estimate can be quickly generated. You can even get software programmes that ask you a few sensible questions and then generate cost estimates for you. I wonder how they estimated the cost of the program that estimates costs?

In some cases an organization might calculate from historical data costs associated with a range of projects and graph those costs. For each new project the estimator estimates complexity and size and deduces a first-order costing.

The estimator has some scale for size and complexity which might be quite rough and ready. He can translate his feelings about the project into a couple of numbers and then check his chart for an estimate. This is really a simplified, graphical version of metrics.

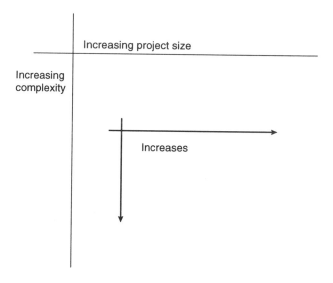

*Relating size and complexity to project costs*

*A project cost model*

A **project cost model** defines the major elements of typical projects and associates percentage of cost with each element. If this is based on historical data, the estimator can use the model to help arrive at a reasonably accurate model if one or more elements are known or reasonably accurately estimated.

**Table 2.1** A typical project cost model

| Phase number | Description of phase | % effort or % cost |
|---|---|---|
| 1 | Definition | 5 |
| 2 | Feasibility | 5 |
| 3 | Initial design | 10 |
| 4 | Detailed design | 20 |
| 5 | Execution | 45 |
| 6 | Commissioning | 10 |
| 7 | Post-project | 5 |

So if you know traditionally how your projects break down in terms like those above and you know – or have a good shot at estimating – one of the phases, you can extrapolate to a full estimate quickly and cheaply. In the example given above the programme-management team have deduced that the cost of the typical project breaks down in the way shown in the table. If you find that the definition phase has cost £10 000, you can be reasonably sure that the overall project will cost £10 000 × 100/5 = £200 000.

The quickest and cheapest method of estimating I ever saw was used by an Iranian road builder. He lived in Tehran and ran a company bidding for and carrying out government road contracts. In the UK this would have meant employing a huge estimating department who would plough through the detailed specifications and drawing of the proposed road and then come up with a price for submission to the client.

There was no such estimating team, he simply weighed the specification. He had a price per kilo of book and sent that in as his estimate. Cheap, quick and if you want to do any cooking you have a set of scales in the office. In those days there was the merest hint of a suggestion that certain government officials could be persuaded to adjust prices later on in exchange for the odd gift like a house in Cyprus but that has probably all changed since the revolution. Now it will be a house in Dubai.

*Types of cost*

You can have many types of costs but there are two very fundamental types of cost recognized by most project-management software packages. These two are known as resource costs and variable costs. You may have assigned resources to tasks and entered the cost of each of these resources. Perhaps four engineers are assigned to an inspection task which has a duration of two days. Assume that each engineer costs £55 per hour. Your project-management software works out a variable cost of 4 × 2 × 8 × £55 = £3520. This assumes eight working hours per day. As you add or take away resources or change durations this cost varies, hence its remarkable name.

Now we turn to fixed costs. In addition to the resource-dependent variable costs, fixed costs can be directly applied to tasks. Continuing the above example, there might be laboratory costs of £1000 which do not depend on the duration of the task nor how many resources are involved. This would be treated as a fixed cost. Building a garden wall would have the fixed costs of design, buying the land, bricks and a gate. The variable costs would include the bricklayers' and labourers' wages.

Most project-management software packages will calculate both fixed and variable costs to arrive at a cost estimate for each task. They will summate the tasks within a heading to arrive at a cost for each section of work until arriving at a total project cost estimate. You can examine the rate of expenditure planned by examining a histogram in cumulative form which is known to all right-thinking project managers as a cash flow curve.

## Influence versus commitment

Here is a thought-provoking graph showing influence versus commitment for which I express my grateful thanks to a supporter of project-management techniques within BP: David Arulananatham, he of the excessively long name and highly perceptive brain, who introduced me to this idea.

The idea is that as time moves forward through a project your ability to influence the project drops away quite rapidly. At the beginning you have a million possible ways of achieving your goals and some pretty hairy ideas get tabled. Very little money has actually been spent to date.

Fig. 2.30

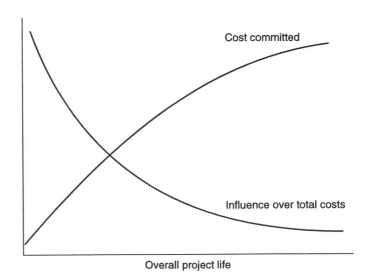

Eventually the selected approach gets designed and the shape is firming up. A little more money has been committed but it is still small bananas. By the time you have contractors milling about with drills and welding torches the really big money is being spent but your ability to influence the overall project costs has diminished to almost zero. You could save a little money by printing more text on each page but this is only playing with the margins. As time passes the spend rate increases, the total amount of money committed increases but your ability to influence the total cost decreases.

# Contingency

As estimates of cost are inherently difficult to produce accurately it is often sensible to include a contingency. It is extremely sensible if you are responsible for bringing the project in to a budget as you will almost certainly need some elbow room. A contingency is an amount of money set aside to deal with the unforeseen. Some organizations formalize the calculation of contingency sums by using a percentage of total costs. The percentage may vary as the project passes through its life cycle. Perhaps a 20% contingency is made in the initial feasibility stage but this is reduced to 10% when the design is finalized.

You can get more or less sophisticated when working out a sensible contingency but there are two much more important political factors at work here.

### Contingency control

Everyone regards the contingency as their own. Don't you? Control over contingency is a common source of conflict as everyone takes for granted their right to spend the contingency sum exactly when they feel like it.

The project team usually begin to feel like it after the first two or three days when the first unforeseen event becomes visible. With the quickness of hand that deceives the eye they'll be diving for that contingency to save the day. It might easily slip their minds that they may need this sum of money to save another day in the future.

Some organizations elect a contingency controller or even a committee whose role is to approve allocations out of the contingency sum. In such environments, managers of phases or parts of the project must apply for release of contingency to the relevant authority. This makes the use of contingency sums a management matter and allows for a proper balance to be achieved. It makes people think twice before reaching for the contingency lifebelt.

By making the contingency not too easy to get to, this approach makes people search for another easier route. Project managers, like water, search for the route of least resistance.

*Contingency location*

Now where are you going to put this contingency? Do you want it to be displayed:

- On your lapel alongside your Train Spotter Club badge or
- Up your sleeve with your elbow room?

Some organizations show their contingencies openly. At the end of the last page of the last section of the detailed estimate for the project are a few lines like this:

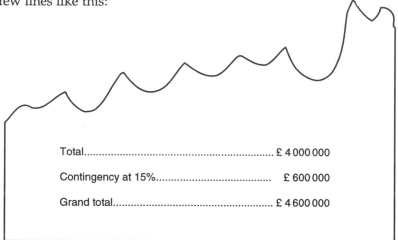

| | |
|---|---|
| Total.................................................................. | £ 4 000 000 |
| Contingency at 15%........................................ | £ 600 000 |
| Grand total...................................................... | £ 4 600 000 |

It makes you think that it was a bit of a waste of time counting all those door hinges and self-tapping screws that cost 0.02 pence each. The contingency is there for all to see as one great lump of cash.

Some organizations hide the sums within other tasks or budget items. If you have a standard company rate for welding chastity belts this might include an allowance for contingency. In the unlikely event of the organization wishing to price a large order, every price, every rate might include a hidden amount for contingency. There is no bald explanation of the sum set aside.

The company culture will often decide which approach is most appropriate. Organizations with internal projects generally are happy to show and manage their contingencies. Firms that bid for work and send those bids in to their clients are generally much happier to hide their secrets. Politics Rule, OK.

## Cost monitoring

It is a bit of a bone of contention with me to separate cost control from cost monitoring. This is one of those situations where the words mislead people completely. Other situations where words mislead people completely include 'Of course I'll love you in the morning' and 'the cheque's in the post'.

> The third phrase of this set is not publishable.

**Cost monitoring** is about finding out what money has actually been spent. This is dead easy as long you have considerable patience or are not in a hurry. It becomes a very challenging affair indeed if you insist on finding out quickly what money is actually being spent. 'Quickly' here means in time to do something about it.

You can find out what money has gone out of the window very easily in due course but this will not be useful information in a year's time. You need to know much more rapidly than that and, probably, much more rapidly than the accounts department can tell you.

**Cost control** is steering the costs so that they stay on course. Trying to control without monitoring is like driving with your eyes firmly shut. Monitoring without control is sitting in the back seat and watching the world go by without saying a word. Sitting in the back seat shouting advice is called back-seat driving and has nothing to do with project management.

Do you know what you are responsible for: monitoring, control or both? How can anyone sell a cost-control software package?

*Problems with cost-monitoring information*

There are a number of problems that come to light when you try to gather cost-monitoring information. The obvious problem is that the numbers are too big. I don't mean they are displayed in 60-point Helvetica, I do mean that the cost of work done to date is invariably too great. This is a problem you seriously want to know about but there are significant problems in actually getting to know anything useful. Some of these problems are explained below.

The type of data
This problem is to do with the form of the data you get. Your accounts department have these wonderful and sophisticated accounts packages that rely on things like 'supplier codes' and 'client codes'. The idea is that every invoice received from a supplier is collected under the supplier's code. Once collected in this way, they are available for inspection. People can ask questions like: How much work has A&B&C Soup Canners done for us this year? How much do we owe this company? Should we pay them something or can we make them sweat a little longer?

This is very helpful for the accounts people but not for projects people. You may not wish to sort through a pile of supplier's listings to find the few invoices that relate to your project. You might prefer to see the list of invoices relating to your job – then you would know how much each firm thinks they have done and deserves to be paid for in the last month. If you can get this information you are beginning to monitor costs.

### Lies, damned lies and invoices

Your contractors may, unhelpfully, submit invoices when they finish the work. Projects would be so much easier if contractors did not insist on being paid for their work. These invoices might cover one or two months' work. Contractors will generally warm to the idea of being paid on a monthly basis as this will improve their cash flow but it may mean extra work for them to do. Each month they have to value the work done to date, deduct the amount invoiced prior to that month and invoice the bit left over.

Building and engineering companies are used to doing this and have special rules to make the process work fairly. They have a system called interim valuations which allows for completed work, uncompleted work, materials delivered but not yet fixed and defects in completed work that need fixing. There is even a complex formula for dealing with inflation.

It is in the nature of life, the universe and everything that the contractor will lie his head off to build the monthly amount up and you will lie your head off to keep it down. The contractor wants to keep his bank manager happy and you want to keep the contractor on his toes. Overpaying a contractor has the magical effect of slipping your project way down his in-tray, well behind other the jobs where he has to chase the money.

### Internal staff

Some organizations deal predominantly with their own staff's time. In these cases most work on the projects is done by the team of experts employed by the company for their stunning brilliance. This is fine and dandy as long as you can get a feel for how much time these people are charging to your project. Perhaps they submit timesheets or some other form of record to their bosses and it might be good if you get an idea of the bits that concern you.

Some organizations plan and monitor in terms of person hours and this can be a very good budgetary control system. Some use a mixture of person hours and money so that the budget is in terms of: £300 000 and 500 engineer hours. The organizations I have seen that do this lump all people together as if they cost the same for these purposes. They are not lumped together as if they cost the same when it comes to the wage bill.

You might find this hard to believe but there are people who spend whole afternoons playing Doom, a game they found on their PC's hard disk. Not only that, having been carried into a make-believe land they make-believe their timesheets by recording the lost four hours against one of your projects.

You may feel that you should use the Auto Lazer Rapid Fire Cannon with optional chainsaw attachment on these people as long as you have enough ammunition and adequate armour. First you must know they are trying to pull a fast one. People facing a timesheet chew the ends of their pencils and ask themselves deep and philosophical questions like 'Where did the week go?' and 'Which project manager won't notice if I charge a few hours to their job number?'

You may not mind very much if everyone dumps their lost hours on you but you will feel a tad resentful if you get hauled over a carpet because your project went miles over budget. So you could probably benefit from a system where actual costs from contractors and actual time spent by internal staff get passed before you on their journey to your project's actual costs.

There are other forms of costs apart from external contractors who submit invoices each month and the internal workforce. Do you have  heavy equipment and if so how is that charged to your project? Are there bulk materials being delivered to the factory which include some destined for your project and how is that to be charged? Of such things great projects are not made. Of such things well-respected project managers certainly are made.

Excuses like: 'I know we went 50% over budget but somehow the whole company's Christmas booze order got charged to my project' is not going to help very much.

## Speed

The next problem to face up to is speed. As I mentioned before, a common and recurring problem is the speed at which cost-monitoring data gets back to the project manager. It may be fine for a production manager to get information back on her baked-bean production process because the chances are she is still baking beans and can make good use of the information.

If the information about the project takes a long time to get back to the project manager the chances are that she has moved on to a different part of the project and has no scope for correction or control. If the design process is running over budget, she can do nothing about the design phase if the product has been prototyped, tested and is being shipped before she finds out the design costs.

I showed a graph a few pages back about the way that the influence you have over the product reduces over its life cycle – the later in the project the less you can do to steer the project to influence cost. It is also true that the deeper you get into a project the less opportunity and the less time you have to bring about improvements in the way the project is being run. Your opportunities on the last day of the project are very limited indeed. The choice is run away or face the music.

This all emphasizes the point that cost-monitoring data has to arrive on your desk quickly to be of any use.

*Studying the information*

So let's assume that you are now getting cost-monitoring data (smack hands: I nearly wrote cost-control data) that is appropriate, timely and accurate. All you need to do now is analyse it and take sensible action. Sensible action could easily mean donning a poncho and retiring to Brazil for a few years but hold on for a moment, there may be things we can do. Let's graph the data for a start.

**Cash-flow diagrams** indicate the expected rate of expenditure over the life of the project. This is normally achieved by building up a cost for each task so that the planning software can summate the costs expected to be spent on each day of the project.

To help with this aim, the project-management software allows the operator to specify at which point within a task costs are actually incurred. Some costs are incurred on the first day of the task, some on the last day and some are spread evenly across every day of the task.

With this information plus the timing of the task within the project you can see that anyone can add up all the money due to be spent on each day. Adding each day to the day before gives you a cumulative view of this planned expenditure. You only need a computer to perform this task because it is more accurate than you and less easily bored.

It is useful when working with costs to select a point when costs are assigned. Choices include:

**Committed:**    costs are deemed incurred when instructions to do work are issued.

**Work related:**    costs are deemed incurred when the actual work is done.

**Invoiced:**    costs do not count until the invoices are received.

**Paid:**    costs are deemed incurred when payments are made.

Consistency is extremely valuable when dealing with costs. If you add apples and pears you might get a lemon.

In these terms you also need to think about any capital that is going to be invested in the project. As well as the ongoing payments for the labour and raw materials going into the project, you may have big sums of money going to pay for expensive computers, buildings or other big items of investment. How do these get charged to your project?

The results of these calculations is a cash-flow curve showing day by day (or week by week) how the expenditure adds up.

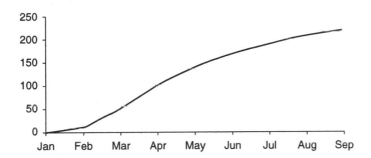

*Cash-flow curve showing anticipated expenditure in thousands of pounds*

This is a simple one-line graph assuming that tasks all begin at their earliest possible times. A second line can be drawn assuming that all tasks are delayed until their latest possible start and finish dates and the combination of these two graphs is called a cost envelope.

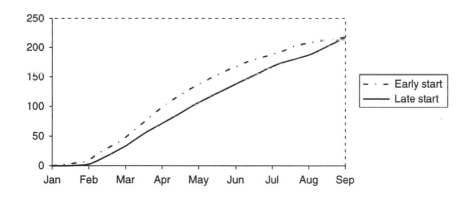

*Cash-flow envelope showing anticipated expenditure in thousands of pounds*

*Project Management Demystified* (this book's elder brother) has a whole section looking at earned value analysis which is a fine way of comparing three things:

- the amount of money you should have spent to date;
- the value of the work you have actually done to date;
- the amount of money you have actually spent to date.

Earned value analysis (EVA), which is also known as C/SCSC or C-Spec, is designed to provide a measure of progress in terms that fairly compare actual achievement with planned. It is not appropriate that this book cover EVA in detail but an understanding of its concepts will help those who wish to research the area.

EVA allows a three-line graph to be prepared which summarizes the actual achievement and actual expenditure on the project in comparison with planned. EVA uses the early start line from the cash-flow diagram as its basis just as it has been described, except that this line is called the budgeted cost of work scheduled (BCWS).

To measure progress it is necessary to physically measure the actual work done or to estimate work done as accurately as possible. Using the rates for doing work that form the basis of the cost estimates and the actual amount of work done gives the value of work actually done and this line is drawn alongside the cash-flow curve. This is called budgeted cost of work performed (BCWP). This second line shows the amount of work done in cash terms on exactly the same basis as the original estimate.

A third line shows actual expenditure and is known as the actual cost of work performed (ACWP). With a little training anyone can interpret these three lines to arrive at a reasonable view of a project's status.

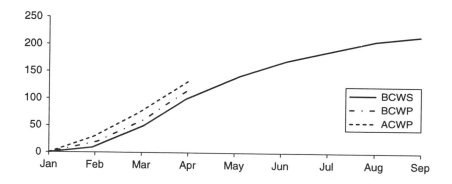

*Earned value analysis graph*

In the example above, the project's status is shown as at 1 April and it can be seen that the project is ahead of schedule but costs are high, even allowing for the extra value of work that has been achieved. Getting these three on one bit of paper takes some doing but gives a good picture of how the job is going. If earned value analysis is too complex or inappropriate for you, you might just draw a cash-flow curve on the wall showing the planned expenditure over the project. You can then draw, each week or month, the actual spend to date. You may make some deductions from this graph but you must be more careful than a mating porcupine as it is very easy to draw incorrect deductions from such a graph.

You might show that you are saving fortunes – actual spend to date being lower than planned. But does this mean that nothing much has actually happened and you are really way behind schedule?

You might show a huge overspend but this could mean that the job got off to a flying start and is miles ahead of schedule. It might mean that you have mistakenly built the wall in the wrong place and had to dismantle it. You are now behind schedule and over cost and it really is time to get that poncho out.

*Control*

So now we have got to a stage where you have collected all the cost-monitoring data, drawn pictures of it, drawn conclusions from it and decided whether the project is in good shape or needs some fitness training. It is time for some cost control. This is where you turn on that remarkable brain of yours and decide what to do. This is where you begin to control the project. This is where most books run out of suggestions and ideas. Including this one.

I'm sorry but at least I'm honest. Ninety-nine per cent of books mix cost monitoring and control together and spend 99.99% talking about monitoring. Control is such a personal, environmental thing that I am not sure anyone can tell you what to do. The whole deal about what is commonly called cost control is in finding out what the situation truly is. That's where systems help. Doing something about it is pretty much down to you.

Hang on a minute, I suppose there are some things you can do. Let's look at some of the things you can do but be warned, this is going to be pretty simple stuff:

Cut something
You find something within the remaining part of project you can get away without doing and without making a big issue of it. Less work means less cost.

De-scope the project
This is management speak for cutting something out of the project altogether. The difference is that you do make an issue of out it and get the

client to agree to the cuts. This is much easier if the 'client' works for the same firm as you.

Use the contingency
You can dip into that contingency or apply to whoever controls it for some extra dosh. Well, that is what it is there for. Some element has gone over costs: to pay the bills you need some money out of the contingency or you will be forced to make savings from some other element.

Revise the project estimate
You could go back to the client and ask for more money to be made available. The result of such a question will depend to a large degree on your relationship with the client. If you have a fixed-price contract with another company she would laugh in your face at your request. If the client is part of your organization the client may agree and then laugh behind your back. If you have a 'cost plus' form of contract with your client they will cough up the extra money and you will laugh up your sleeve as the project budget and your company's profits go up.

Get something done cheaper
Perhaps you can find a way of doing everything within the scope of the project but doing some things cheaper. Seeing as you found out about the overspend nice and early and you can add this information to your budget, which is carefully broken down into categories, you are in an ideal position to look at the project for savings. It is amazing how a careful scrutiny under pressure finds some areas to cut costs.

Maybe you can substitute steel taps for gold ones. Could a cheaper carpet do? Once again it might be time to talk to your client or manager before making decisions. It is easier to go with cap in hand and make her aware of the overspend on some early work if you are armed with some ideas for bringing the project back on line.

You can extol the benefits of your excellent project-management techniques which brought the problem to light in time to fix it. It may be bad news, but it is not bad news that came too late.

# *Project-management tools in the programme-management environment*

## *Do we need one?*

I guess the first question to ask about tools for programme management is: do we need one? Software people, and particularly those wearing smart suits, carrying mobile phones and driving Vauxhall Cavaliers, are not reliable sources of information on this topic. Even if they don't actually want to sell you something, they tend to come from the 'It can be done, therefore it should be done' school of computing.

There are some sensible factors to be considered when you start to think about programme management. Step one should be: take a long hard look at the organization. You really don't want to start with computer tools. You might easily end up making the same mistakes you have been making for years rather more quickly than before. You might easily find yourself making some new mistakes.

Try to figure out what you want to achieve and then to find sensible systems to manage the process you design. The tools you need might be pre-printed forms, a wallchart with flashing lights, a Lego barchart drawing kit or a bit of software designed for the job.

You might need software tools to support your programme-management techniques for these reasons:

- It is easy to encourage or even enforce standards through a software system. It is simpler to ensure that everyone uses the same milestones on their plans and that every plan shows the design work being signed off by the board before manufacturing begins. It is easier to get people to plan in a consistent and predictable way with a systematic approach and a software system will underpin this.
- It should not be complicated to extract from the mass of data the information each manager needs. Each manager's requirements will be different: project managers want details of their projects, departmental managers want to know what is going on within their teams, programme managers want an overview of all projects, individual resources want to see what they are supposed to be doing. This kind of manipulation of data is what computers are good at doing.
- There is going to be a lot of information. You might easily have three or four thousand tasks in a plan mixed in with resources, calendars, costs and baselines. It all gets a bit voluminous – not complex but overwhelming. Computers, especially big computers, are good at dealing with loads of numbers.

## Classes of tools

When we think about software tools for programme management there are different approaches. The three basic classes are:

### Stand-alone systems

Here a simple, easy-to-use, convenient software package is purchased and installed. Most people can fairly quickly get the hang of such a system and turn out neat-looking barcharts all day long. You could even sprinkle a few copies of the system around the organization. The benefits are that you can encourage a little consistency in planning and you get people planning in the first place. You do not get an overview across many projects. You do not get a feedback mechanism. Each plan is an external model of one project which is useful to that project but not to the enterprise as a whole.

Such systems are cheap – expect to pay from £100 to £1 000. Popular names include Microsoft *Project*, *SuperProject*, *PMW*, *Timeline*, *Instaplan* and *Powerproject*. These are systems you could use once a week. They are the Mini Metros and Ford Mondeos of the project-management world.

### Multi-user systems

This is another ball park all together. The only things that multi-user project-management systems have in common with stand-alone systems is:

- their basis of critical path
- the name tag of 'project management'

Multi-user systems run over a multi-user operating system like UNIX or VMS. Lurking at the centre of a web is a central processor that knows about critical path and budgets and so on. Hooked up to this web are terminals of varying degrees of intelligence which allow you to enter and update details of your project.

Your details are added to everyone else's details in the pile of data held at web central. This means that anyone with the necessary authority and the enthusiasm to overcome the hurdles set up by the necessarily complex multi-user systems can get as many overviews of the data as they want. Barcharts showing summaries of all the work and histograms showing the demand for resources and skills across all projects are available.

Such systems require teams to maintain and support them and cost loads of money. Typically the users spend a good deal of time on the systems and get to know them well. They are the furniture vans and tank transporters of the project-management world. Names like *Artemis* from Lucas, PX from PSDI, *Cascade* from Mantix and *Panorama* spring to mind. Prices go from £25 000 up into the stratosphere.

It often happens that an organization purchases a site licence to use such a system allowing for its 50 project managers and planners. The system is so hard to use that only a few actually get into the system and make it sing for its supper and these one or two people quickly become almost full-time planners, doing the planning for everyone else. This is OK except that the cost of 48 user licences has been wasted.

*Combined systems*

You can get close to what might be the best of both worlds for some organizations. You give the planners a simple front end in the shape of a stand-alone system and use a more powerful tool in the project office to merge plans together. This is a good plan and works well in some environments.

You take everyone's plans in from the stand-alone systems, add them together with the heavyweight software and examine the conflicts. Resolving said conflicts means changing plans and these changes need to be reflected in those stand-alone plans. You probably have to tell the planners to change their plans to bring them into line with the master plan.

There are problems of data compatibility and consistency here and the data transfer is generally one way only. This means that once the plans are all absorbed and collated, once decisions are taken about the priorities of the projects, these decision have to be communicated verbally and the stand-alone plans modified to suit. The programme manager may call on Jack and say 'We're going to have to delay the nose cone project testing until 27 February because of the pressure of demand on the test rigs. Can you change your plan to suit?' This is not a very automatic system and mistakes are likely to happen.

# Using the tool

When it comes to planning multiple projects in industry in the UK I have identified four stages in the planning process. Actually I have identified five but few people seem to bother with the fifth.

**Planning:** the creation and editing of individual project plans either on stand-alone PCs, on PCs hooked up over a network or on terminals connected to a minicomputer.

**Transmission:** physical movement of project-planning files from the computer on which the plans are built and maintained to the computer on which they are to be consolidated.

**Consolidation:** the merging of many plans into one model.

**Decision making and reporting:** the examination of that master plan, the identification of problems and the communication of those problems. The communication of the solutions.

Below, a few case studies are briefly summarized explaining how some well-known organizations coped with these four stages. Later there is a mention of the fifth stage – the feedback mechanism which I see as being critical.

## Case studies

These are brief studies of well-known companies who have addressed the programme-management issues in their own way. Each organization is briefly outlined and their approach explained. My idea in including these notes is neither to criticize nor admire these solutions – it is to explain the way they work so that you might be given a wider range of options to consider for yourself. I have made these observations of these organizations over a number of years and the organizations may have moved on.

### IBM Hursley Park

IBM Hursley Park run a number of software and hardware development projects in their high-tech laboratories. People tend to be white-coated, high-tech, well-qualified engineers who are very computer literate and capable.

On each project the planners work with their own copy of SuperProject to plan their research and development projects. These plans get transmitted fortnightly via floppynet (copied onto floppy and carried over to another computer) to a project office where they are consolidated into a complete and very large SuperProject plan via a SuperProject macro written in-house. The project office people can inspect overall histograms and summary barcharts. Decisions are taken and communicated verbally back to the planners.

### BT Ipswich

Here many simultaneous projects proceed through a set pattern of design, specification, programming, testing and installation. The projects are nearly all software-development projects where a new version of an existing program is created to run some telephone switching gear.

This organization runs a typical matrix with resource managers looking after a department dealing with each stage of the process and project managers looking after each project running across the matrix. It is the resource managers who are departmental heads and therefore functional managers and who plan their workload.

They do not plan projects but mostly plan unlinked tasks passing through their department with *Project Manager's Workbench* (*PMW*). This data gets transmitted weekly via their local area network before consolidation in the heavyweight UNIX-based *Cascade* in the project office. The planning team can add links and check for conflicts between projects.

As the resource managers plan their workload it is reasonable to assume that there will not be too many resource conflicts. There are, however, likely to be conflicts within the projects as tasks await resources or are planned to happen illogically early. This is an unusual approach and a fuller case study is included on page 208.

### National Power (NP) at Drax

This is a project where a programme means a very large project comprising many connected projects. Here large scrubbers, mills, chimneys, ducts and other items combine to make a huge engineering plant operational.

Initially NP had many planners looking after phases of the work – design, construction, commission – across the many elements of the programme which is really one huge project. This was done on multi-user *Artemis* so consolidation was immediate and automatic. When resource allocations were made they could check for resource needs and also maintain links between elements.

For example, commissioning one pump might allow the commissioning of another vessel to proceed. So here a heavyweight system brings together many plans with many planners working away at connected terminals.

Interestingly, today, NP mostly subcontract the work and therefore use very few in-house resources. Because of this they have no need to plan resource requirements as this is the concern of the subcontractors. NP hence switched to simple presentational bar-charts on *Powerproject*.

### Westminster City Council (WCC)

WCC run a wide range of projects for the benefit of local residents. These include road improvements, landscaping, foot bridges and so on.

In WCC the whole team was regarded as a pool available to perform the skills they had. Everyone was trained to gain project-management skills. The aim was that a typical engineer might be project manager on one job, assistant project manager on another and a resource doing landscape design on a third. Landscape design would be the person's specialist skill.

Time planning is done by everyone – architects, engineers, landscape designers – all manually. Normally plans are entered into the heavyweight system *Trakstar* by a small team of 'operators' who know the system, know planning but need not know the work. The planners pass the information over to the operators who enter the data.

Again, consolidation is performed immediately by the software so that the operators can print and distribute reports.

### Laserscan

In this high-tech environment the company work with electronic maps. This is much cleverer than simply a graphic image of the maps. Electronic maps know what each element on the map represents. It can sort out rivers from roads, houses from hills and is therefore much more useful than a mere drawing. You can ask for grape-growing zones and it will find land with the right slope, facing the right way and accessible by track. Laserscan's projects involve digital mapping for various organizations including assorted national defence authorities.

The planners plan their work manually and pass the data over to the small team in the 'project office' who enter details into *Milestones etc.* – a small, simple planning system which is almost a barchart drawing tool. Plans are separate so conflicts are viewed by comparing plans visually. Plans are all displayed in the project office which is more like a war room than anything else. Managers can stroll along to see where each project is up to and what remains to be done. Little resource planning goes on here which is surprising as most of the resources are very highly-paid software people. The organization finds its needs best satisfied with a simple approach to planning.

### Wedgewood

This firm is a household name and their projects involve the introduction of new products. When I visited there were 102 projects, market. Projects began with a possible gap in the market and ended as the new range went into production. The projects followed a predictable path through various separate functional departments. The jobs were grouped into three programmes of work, each with a project manager and a planner. They had three copies of *SuperProject*. The three separate plans each covered a group of projects. Consolidation within each programme was immediate as the software was able to 'link' the thirty-odd plans together. This meant distribution of three plans to resource managers who were trying to meet the demands or complain if they foresaw overloads.

### Research Machines

RM is a high-tech firm making mostly British computers. Their projects are new products which might be hardware, software or network products. The company has departments specializing certain types of work – printed circuit-board design, software development and networking products, for example – and the many projects vie for the resources within these departments. The more general management form teams often with part-time members to manage these projects.

They have a home-brewed local-area-network based software system. Anyone can add in a new project to the organization-wide programme using a library of typical projects or create one from scratch. Plans are normally simple but loaded with resources. Consolidation is immediate via the network and purpose-built software. Each terminal can display a project, summaries of groups of projects and demand for each resource.

## NRA

The Thames Region of the NRA are involved in many development and maintenance projects. Most involve a lengthy design and planning consent stage before commissioning and executing the work on site. We are talking here about mending a lock, renewing a weir or installing a fish-protection scheme.

Loads of planners worked with *SuperProject* to plan their river-maintenance projects and transmitted their plans weekly via floppy-net (using coloured disks sent by post) to head office. Plans were consolidated via a purpose-built database program permitting the analysis of histograms and summary barcharts. One could also list the projects in various sequences, for example, in order of depth of trouble. The system was designed for timesheet data to be fed back into the purpose-built software which tried to complete the circle by updating plans before they got sent back out via mail (on different coloured disks). Implementational problems figured significantly in this system.

## Choosing a software concept

Long before choosing a tool, you must first choose your approach. In the magazines and at the shows is a very wide range of software tools which can easily be categorized once you know where you want to go. You cannot tell from the adverts and sometimes you cannot tell much from the brochures. These systems follow certain fundamental approaches to multi-project management and it is the approach you must think about first. Two of the fundamental features of any programme-management software tool are going to be:

plan individual projects and
ose projects together.

to be two approaches to doing this and, so that you dis-
men by a searching question, here are those two ideas.
as two 'sub-approaches' so here are three approaches in all
value you get with this book.

### Develop plan for each project and merge them together

Here you use a small, cheap PC system like Microsoft *Project*, *SuperProject*, *PMW*, *Timeline* and so on and give each project manager a copy of the software. Each project manager plans his own project, hopefully sticking to the standards laid out in the company's planning procedures. By this means you will have 60 plans for the 60 ongoing projects. Then you collect these together onto one machine.

Then you merge or link these plans together so that you can deal with the overall, cross-project resource demands and get reports dealing with like aspects of the many projects. You can simply merge the whole lot into one big plan. When I say big I should say huge because this could be three or four thousand tasks. The machine you need to achieve this would probably need a good supply of memory – random access memory (RAM) in this case.

Assuming issues like task numbering, task identification and resource names have all been dealt with in a simple, non-confusing and consistent way, this merge operation should all work fine. You might at this point add in any logical links between the various projects – perhaps one project creates a subassembly that another project awaits. The result is a huge plan from which you can extract parts and examine overall resource predictions.

Sometimes this merging is achieved by regarding the many projects as subprojects which may prevent logical links between tasks in different projects. Sometimes this combining of projects is achieved by **linking** which means that all plans are there sharing memory space but are not really in one big plan. They are and are not one large plan. You probably cannot create logical links between linked plans and you may not be able to select to see a group of tasks down the matrix – can you select to see all 'site installation' tasks in all the linked projects?

After a consolidation process you may not be able to **unmerge** plans. Once they are together in one plan it is likely that you will adjust and polish the plan changing start dates and end dates to improve the balance of resource demands. You may not be able to send these new plans back out to their respective managers. If you can't you will have to tell (agree with) everyone what changes they should make to their plan to bring it in line with the master schedule. Linked plans are generally easier to unlink and send back in their modified form.

A nice thing about this basic strategy is that everyone plans their own work which happens to be something I strongly believe in.

*Develop one large plan for the overall workload on a single computer*

In this approach you commit one computer to planning the work. This probably means committing one or two planners as well as a computer to do the planning work for the individual project managers. These full-time planners can use one of the heavyweight planning systems which are generally as user friendly as a cornered rat but, since they spend their working days with the system, they can live with it.

These systems allow very large plans to be built and are bewilderingly powerful. They tend to eat memory in terms of both RAM and hard disk space and need big computers to run on. Often they do not run at all, or rather slowly, in the PC operating environments like DOS and Windows. You may have to move up to UNIX, VMS or something larger like that.

As there are a couple of full-time planners you can expect a reasonable level of consistency between the various parts of the plan. They should be able to structure the data to allow for the multi-project environment and for the ability to extract data in whatever form you dream up. Expertise is not needed by the project managers nor the senior management as they simply ask the lads for the information they want, sit back and wait.

On the other hand you are reliant on the 'experts' you employ who can become rather central to the company's future. If they are weak, or decide to blackmail the company or get tempted away, you're in bother. Also, your project managers do not plan their own work, they get the experts to do it for them. The attitude can quickly spread where people say 'JD wants a barchart for the Spaghetti job, can you knock something up to keep him happy?'

This is clearly not useful and productive planning – it has become a fop to satisfy someone called JD and if that happens to be you, you're in trouble.

*Develop a multi-user system of planning*

Here you install a multi-user system for project planning. You might install a PC local area network (LAN) or buy in a minicomputer and install terminals around the place. You can even combine LAN with a minicomputer system. Everyone does their own planning on their own terminal but everything is transferred electronically to the centre where an expert planner or group of planners examines the cross-project resource needs and extracts reports for senior management.

The advantages here are that everyone is planning their own work but they are fighting with a very heavyweight system like Lucas' *Artemis* or Mantix' *Cascade* or PSDI's *PX* which does require significant training for each user.

The data is electronically gathered and, as electricity travels at the speed of light, not much else happens in the time it takes the data to get back to data central. However, these systems are very expensive in terms of both initial purchase, installation and training. You will probably have a small team of people to look after the system who are not involved in any project-productive work.

These systems involve a big bang approach. After a careful evaluation period the user places the order and stands back as the huge system is installed and tested. The implementation process takes some time – a year would not be unusual for a large organization – and training goes on for ever. It is a real commitment.

## Choosing project-management software: a strategy

This is an interview taken from *Project Manager Today* which discussed the problems of choosing a project-management software product. The actual product selected is not as important as the way in which it was selected. I found GPT to be very typical of the state in which most organizations find themselves when they start to think seriously about project and programme management. GPT's approach to the problem was exemplary. I am grateful to GPT and *Project Manager Today* for their contribution.

GPT abbreviates GEC Plessey Telecommunications Limited and is the name of a large company whose products are involved in many phone calls made throughout the world.

Formed in 1988 GPT invests large sums of money researching and developing into telecommunications as well as carrying out projects both large and small for their clients. GPT's clients include BT and Mercury and phone companies around the world. When you make a phone call you are almost certainly relying on GPT products.

GPT was formed to create an 'integrated and cohesive business' and now has some 4 000 engineers involved in a wide range of projects both large and small.

The company took a policy decision to have harmonisation across its many projects and, in common with many other companies, wanted to improve its management of projects.

At the time of the merger Plessey had a project-management user group but little co-ordination and a few specialists in various project-management systems. GEC were carrying out a feasibility study trying to overcome the unfriendliness of weight main frame system in use at that time.

Interviewed were two of GPT's project-management te Coventry facility. Keith Oakley is Development, Visi Control Executive and Martin Harris is the Programme M

Both of these two were involved in a lengthy project to select a company standard project-management software package. They approached the problem with a classic engineering approach.

*PMT:* Let's start at the stage where the company took a decision to harmonise its project-management approach – why was this decision taken?

*KO:* We saw a number of advantages and disadvantages to a common standard project management package. On the positive side we saw:

    staff and project mobility;
    common training;
    support via an internal user group;
    purchasing power;
    cross fertilization across project-management teams;
    common code of practice; and
    the ability to influence the supplier.

On the other hand some of the problems we had to accept were that project management teams would have little individual choice and that we would become reliant on one supplier.

At this stage we also decided not to develop our own system – this is a general GPT policy – and to have a common hardware platform for common systems.

We did want to distribute the planning tools and have a central system for collecting projects.

*PMT:* You could have chosen to set up a central planning office which would execute the planning function for your project managers.

*KO:* That is not in line with our policy. We want the project-management teams to have their own tools and be able to collect data from many projects together.

*MH:* We should point out that project-management software is planned to form a part of a greater control system called the Engineering Accounting and Project Management System (EAPM).

*PMT:* Was the next stage to establish some sort of requirement?

*KO:* We started this process very early on. We formed a group to establish the User Requirement Specification. This group's task was to establish a precise need but this went on in parallel with the decision about harmonization.

This group talked to 'hands on' people and developed a list of features. Each feature was weighted in terms of importance as we expected not to find every feature in one tool.

There were battles at this stage over planning techniques. We had activity on arrow users on Plantrac who tended to be more mature planners and we had a precedence lobby as well. We eventually choose precedence as there were more available tools and because the world seemed to be going that way.

So we got to a stage where we decided to standardise on one system and knew what we wanted from that package.

Another early task was to find out what project-management systems were in use in the company. We carried out a survey and found that we had about 40 copies of 9 different systems. About half had been purchased and the rest were evaluation copies. No one package seemed to stand out at that stage.

*PMT:* So we next move onto the evaluation stage?

*KO:* Yes. We set up evaluation teams taking people from varying levels within the organisation all of whom were project management types.

We assembled a short list of packages from:

packages in use already;
the market leaders;
other packages that seemed worth examining.

A first look at these packages and discussion with the existing users did eliminate some systems at that stage. Our short list contained nine packages as we entered the next stage.

We asked the vendors to send us evaluation copies of the latest version and, if the vendor thought it wise, we sent our evaluation teams on the first training courses. An evaluation team was a two-person team armed with a standard scoring system.

At this stage the teams were to consider each required feature and evaluate the package. Each package got a score for each feature. An 'A' indicated that the package met the requirement, 'B' indicated adequate, 'C' was inadequate and 'D' meant no good at all.

As you would expect, this did reduce the number of packages. The short list went down from our original nine to four.

Then we switched the teams around so that the second evaluation of each package was carried out by a different team. At this stage we also had learned enough to refine our scoring system a little.

*PMT:* Was any one package standing out by this stage?

*MH:* Not really. The scores were all very close, there were strengths and weaknesses in specific areas but there was no clear leader.

At this stage some elements were emerging as being more important than had been thought. Price was never a major factor as the actual purchase price is not large in comparison with implementation costs and is small in comparison with the value of a project.

Integration had become a larger issue. We saw the importance of the need to be able to integrate data from project managers and this requirement was added to the list. Although it is hard to evaluate we decided to try to measure ease of use as a feature. We also took into account the hardware requirement of the packages.

*KO:* One danger is that users, when specifying their requirements overlook the fact that each extra requirement tends to mean more complexity of operation and therefore less 'user-friendliness'. Functionality is inversely proportional to ease of use.

*PMT:* What did the short list look like at this stage?

*MH:* I think we were down to three – Qwiknet, PMW and SuperProject Expert.

We finally closed in on Qwiknet in January 1989 and came to a commercial agreement with PSDI, the vendor. We issue our licences and every user gets a full copy of the software and documentation.

*PMT:* Then the hard work really begins?

*KO:* Absolutely. Training was the first problem area. We knew we would have to train people in the use of the tool but we seriously over-estimated the level of knowledge of the topic. We set up a system of issuing internal licences to use the software and a user had to attend a training course before getting a licence. We had at that time about 190 target users.

We quickly found the need to teach the techniques behind project management so we had a two-stage course – part one looked at techniques like critical path analysis – this was called PM1 – and part two was a hands-on session with the software.

Naturally the 'methods' teaching related to the use of Qwiknet.

This worked well but a message came through very clearly from the project management teams. The message was that their bosses ought to attend the course as well.

To meet this need we set up a 'PM2' course to give managers an understanding of project-management techniques, trying to explain to them what to expect from their staff. This worked very well but we got the same message back. The bosses thought that **their** bosses should go on the course.

To meet that need we set up a 'PM3' project-management course aimed at senior managers. This was a one-day course split into half a day on techniques and half a day discussion about project management policy.

To date we have run about thirty Qwiknet courses with eight delegates on each, twenty PM1 courses with fifteen delegates on each, ten PM2 courses with fifteen attendees and seven PM3 course each with seven senior managers.

*PMT:* That seems very comprehensive. Did you design the courses internally?

*KO:* No, for the same reason that we didn't write our own project management system we used People & Projects to design and present PM1 and Performance Consultants to design and present PM2 and 3.

*PMT:* How strict are you about the training need?

*KO:* Very. No training equals no licence. Central Engineering ran the project-management software selection project and now issues the licences. There have been 250 to date. Increasingly these licences are for Local Area Networks which allows for easier integration. We still have Project 2 on the main frame and integration is still a stated policy.

*PMT:* How about support – does this happen within GPT?

*MH:* Yes it does. Central Engineering set up a user group which meets every one or two months. This group agrees codes of practice and lets users compare notes. We do invite PSDI, the vendor, to these meetings. They often go away with a wish list of new features.

PSDI does not have a permanent representation within GPT but does have a GPT representative in Mike Gaff.

*PMT:* How have the project-management staff reacted?

*KO:* One thing we found was that the benefits of harmonisation tend to be seen high up the management tree and the disadvantages tend to be experienced by those doing the harmonisation. For example we have to persuade a project manager to adopt a different package which may not be significantly better than the current package. We cannot 'sell' the idea on the basis of the advantages to the planner, the benefits are derived by the senior manager who can easily combine many project plans very easily.

This has led to codes of practice and user groups as we feel that the problems begin with the occasional user. Regular users working in a standardised way clearly makes the whole thing much smoother.

*PMT:* What is the long-term strategy from here on?

*KO:* This harmonisation is only phase one – to create a common project-management and planning system. Phase two is to establish a system for integrating the planning systems with the accounting systems, giving us the capability for Earned Value Analysis, costing and cost monitoring.

Phase three is complete integration giving us multi-project planning and resource allocation.

*PMT:* Now that you have been through the software-selection process what words of advice do you have for others just setting about the same task?

*KO:* Do not overestimate the amount of knowledge within the company in project-management techniques. Check the manual systems that are in use and evaluate their effectiveness. It is important to involve the end users so as to avoid 'selling' the system after it has been selected. Oh yes and watch out for changes in the project-management environment that may cause a change in the requirement.

*MH:* One other point about the selection of a supplier. There are two extremes – the large reliable supplier over whom you have little influence and a smaller, potentially less reliable supplier who will try to meet your needs.

## The phantom project

No – this is not the twilight hour. No nasty project is going to leap out at you from behind a critical-path diagram covered in tattered black clothes, gore and blood and mud and a sickly grin or with scissors instead of hands. The phantom project is the maybe project – the project that you might and might not be doing.

If you are in the business of trying to win work in a competitive marketplace, if you bid for work on a competitive basis or submit proposals for work to your potential clients, if you wait for approval from head office before proceeding, then you have phantom projects. There is a list of jobs that you expect to get, might get, but which are not yet firm.

The problem arrives squarely on your nose when you are supposed to plan the workload for the next few months but don't even know which jobs you are going to be doing. You cannot plan to use all the resources on the firm project workload as that will leave nothing left over for the phantom jobs, some of which you expect to win.

Remember that in the world of programme management we are not talking about one plan with a start and a finish; we are talking about a continuous workload which has no start and no end. New projects arrive onto this ever-moving plan and old projects fall off the back of the plan into history.

You cannot plan to do them all as that would show a huge overrun on resource demands. So how to deal with the phantom projects? Here are some ideas.

### Don't plan them at all

This sounds daft at first but actually works very well in some organizations. You make no provision for any phantom job until one becomes firm. Immediately a new job becomes firm you go into a fairly busy few days replanning the workload to allow for the new project. You knock up a detailed plan for the new job and add it into the schedule for existing projects. It helps if the number of new jobs is small and their demands for resources are not instantaneous. If you get a job that gives you three months' breathing space you don't need to plan before it is firm. It also helps if plans are drawn up of these maybe jobs. You might draw up a plan for the phantom job as a separate individual plan and show off to the potential client your undoubted project-management skills.

Once the job becomes firm, creating a real plan is eased by the use of the existing plan.

### Do different plans

This is likely to be relevant when there is a small number of big projects which will have a big effect on the workload. You produce a plan which covers the situation where the big job does not arrive and another plan where it does. You could continue this a little further and, if you have two big but different phantom jobs, prepare plan A – if job 1 arrives but 2 does not, plan B for the opposite, C for neither and D (lawdy, lawd) for both being won.

When things settle down you drag out the appropriate plan and it becomes official. It gets really hard with just three jobs so that is where you start to play the odds.

Play the odds
This is an approach where you rate each project by the chances of getting it. Firstly you plan each phantom job in the same detail as any other job. If you are 50% confident of winning this particular tender then this job is rated at 50% likely. You can apply such a percentage to each project and in particular its resource demands. A 50% likelihood project might demand only 50% of the resources it would really demand if it became firm. By this technique you can plan all the possible projects even in the near future. By planning two projects each with a 50% likelihood, you are saying that one of them will become real and one will disappear but we don't know which.

It gets more complex when you have 20 possible projects all of different sizes and with different likelihood's but you can see that reasonable estimates can be made of future demands for resources particularly if the projects are similar – runners or repeaters. If your software allows you a way of entering this likelihood percentage once per project, acknowledging a change in odds – like you won the job – is very simple.

## Feedback and the timesheet angle

Using your brilliant mind plus the wonderful interactive, user-friendly, windows-compatible, fourth-generation, artificially-intelligent, client/server, globally-enabled decision-support software you plan the workload in appropriate detail.

| Just practising for a copywriter's job. |
| --- |

Very soon your wonderful plan is ruined because someone does some actual work. How dare they! Not only did they do some work and ruin your plan, they don't quite do the work you planned for them. Because reality has the annoying habit of not fitting in with your plans, the system should loop back with feedback so that people are advised what has been actually happening.

In many organizations the project managers cannot simply measure progress as there is no physical deliverable. Life is simple for the builder who has bricks to count but much harder for computer people, research engineers and technical authors. There may be distance or other problems in measuring actuals.

Hence the timesheet angle. Timesheets input is very common in some environments. It is particularly common in those organizations where most work is done by highly-paid professionals – situations like software development and R&D. Update information is created as everyone enters data into timesheets which is fed back into and updates the project plans. There must be some human intervention here as we all know that eight hours' work on a task does not necessarily mean eight hours' progress!

Timesheets are a pain – they are almost universally loathed. I think this is the case because the people that do the work of filling out the timesheets get no benefit back – they do it because someone has told them to do it. And they can take ages to do. Perhaps people dislike timesheets because they are a reminder of those horrible and belittling clock-in machines where it was assumed that you would leave as early as possible and arrive as late as possible and so a machine would be set up to watch you.

Very often timesheets are seen as a policing system which are used only at random and only to check up on people. This is true in some companies. The timesheets are carefully filled in and collected and then stored for the statutory seven years or whatever it is. No one looks at them or uses them except the odd manager playing detective who decides from time to time to pull at a form or two at random and check that the story fits.

Sometimes timesheets are used to create income. No, I don't mean selling them as great works of fiction, I mean using timesheet data in the invoicing process. You start with 20 highly-paid engineers beavering away at various jobs, keeping their timesheets up to date.

Every timesheet contains person-hours that have been spent on a job for a client and this means costs. Sometimes you simply multiply the person-hours by the agreed rate for the work and there you are – with an invoice to be paid. Or at least the information that goes onto the invoice. Solicitors and accountants use an approach like this. Some timesheets drive the wage calculations and these are the home of many great works of fiction.

In some strange companies there are separate timesheets for work-load planning and costing/invoicing/wages. Where is the sense in that? Bureaucracy has stepped forward and become more efficient at the expense of the efficiency of those who actually do the work.

So timesheets need to be quick and easy to do and should help provide a benefit or at least show a benefit to the people who actually fill out those timesheets. They should understand why the information is useful and what is going to happen with it. Otherwise they get a laugh at first but quickly get bored with the whole thing.

The deal you might one day be able to offer the resources goes like this: I'll plan the workload throughout the whole organization working with the project managers and departmental managers and you'll get a report showing you what you are supposed to be doing over the next few weeks. Each week you do a timesheet and I'll replan the workload working with the project and functional managers and issue you with a new barchart showing the next month.

The deal gets better. You can explain that when someone comes along to ask them to work all night to finish off some job that he didn't plan; when a director demands that you drop what you are doing and collect the client from Heathrow, you can reach for your barchart or timesheet and show, prove and justify the way you spend your time.

Such a deal might get people slightly more keen to fill the damn things out. What can you, the programme planner, do with these timesheets? You can add the information contained in the timesheets to your own knowledge of what has been going on and therefore update the project plan. The original estimates of durations and resource requirements for each task were just that, estimates. Sometimes they are guestimates, sometimes they are guesses. Whatever they were, you now have some new information with which you can update your plan.

---

A guestimate is a cross between a guest and an inmate, therefore a Blackpool B&B resident.

---

You might also have some physical measurements to take: how many gearbox casings are ready; how many leaflets are printed. You might have the project manager's dire warnings – 'This is going at a snail's pace.' You might have the functional manager's reassurance – 'It's going really well.' Put all of these clues together and you can revise your estimates of how long those tasks will take to finish and what resources will be required.

It therefore makes sense not only to ask how long each person has spent on each task, you want to know how much more time they think they will need. Don't bother to ask how far they have got because answers to such questions are unreliable and assume much about the accuracy of your first estimate. Ask how long they think it will take to finish and you'll get a much more realistic estimate which takes less notice of your first shot.

Page 164 shows a layout of a fairly sensible timesheet which you might use.

It makes sense to set up a system to issue these forms on some sensible regular date – every Friday morning might be good.

Whoever sends out these forms might have the job of listing the sort of tasks that the resource is likely to have been working on. Alternatively the resources might have a task list which they can use to refer to when filling out the timesheet. The purpose of these two ideas is to make the language consistent.

Some organizations issue a short-term barchart showing the tasks that should be going on but leaving space for two types of information: the actual work that has been done on the listed tasks and the actual work that has been done on tasks not on the barchart. Everyone is supposed to add the extra tasks, fill in the columns about actual work done and hand these barcharts in on Friday evenings.

# Norfolk & Goode Timesheet

## Name: Fred Guggelheimer

Period Covered From:  18 Sep      to:  25 Sep

Submit by: 26 Sept

| Project number | Task description | Actual hours spent this period | Current estimated total cost of task | Current estimated end date of task | Actual start date | Actual finish date (enter only when task is complete) | Notes (version number of product, delays and problems) |
|---|---|---|---|---|---|---|---|
| 14/42 | Design nosecone | | | | | | |
| Eh? | Design earcone | | | | | | |
| 99 | Design ice cream cone | | | | | | |
| M25 | Design traffic cone | | | | | | |
| | | | | | | | |
| | | | | | | | |
| | | | | | | | |
| | | | | | | | |

When all these timesheets come flooding in from the various depart-ments you, dear planner, might have the job of trying to link these entries with the tasks in your plan. Using consistent task descriptions is going to help you a great deal when you try to decide if Cheryl has been designing the nosecone on the new missile project or really designing a new earcone for the Punch and Judy show.

In an effort to make your life easy, it makes absolute sense to extract the task names from the project plan itself. Maybe your software allows you to build up a timesheet form showing the tasks that the person might be working on and leaving the blank spaces alongside for them to fill in this week's details. The form goes out looking just like the one opposite – it shows the period it covers, when it should be returned, who it is for and the likely tasks.

You should always leave space for the person filling in the timesheet to add some tasks that you did not think of or even know about. During most weeks people are whipped away from their project work to clear up mysteries about long-completed projects or discuss proposals for future projects. They need the space to add in those extra tasks that you had no idea about.

### The lost timesheet

You hardly ever get 100% returns so what do you do about the lost or delayed timesheets? How do you deal with the absence of a few timesheets in your system? One way to help increase the percentage of timesheets submitted is to link the system to payroll – no timesheet, no pay and no bonus. Of such things great motivation systems are created.

The hard way is to assume that if there is no timesheet, no work has been done and therefore no progress can be reported. This is hard and firm but does tend to help get those timesheets in. The next version of the plan shows that during the last week no progress was achieved, the work that was there to be done last week is still there to be done and any project manager worth his salt will be on the phone immediately he sees the new plan. The information is probably untrue as work was probably done, progress was achieved and the resources are now moving onto work on some other stuff.

On being asked, they will quickly reassure the project manager that work is going on but there was a problem and the timesheet didn't get submitted on time. Still, the message got through.

A much kinder way is to assume that work went on in accordance with schedule and that when the timesheets show up they will prove that this is the case. The new plans are issued showing that the work was done to time and shows how the resources can now move on to new work. This will probably also be untrue as those involved most likely did some of the work but didn't quite finish it and have to spend the few hours of the next week finishing off what they should have done last week. This approach is therefore optimistic to the point of being dangerous.

Perhaps you could set up a system where you made assumptions based on other information, phone calls and your own intelligence to update the plan and mark the appropriate tasks in some way:

****No timesheet submitted – estimated****

This is the technique used by the Electricity Board when they visit your house to read the meter, knock on the front door with a sponge and sprint away back to the van. They estimate the reading and give you a little E after the figure.

There is a loose connection between missing timesheets and aggravation. People never like to be the bearer of bad news. When a task is going badly, people at first tend to lie to themselves about how well it is going and report back happily. Later, as things get worse, they stop lying to themselves but go on lying to the project-management team and report back happily. Later still, when things get really bad and the dreaded day is approaching when these lies are all going to come out into the sunshine, they stop reporting all together. Their logic, if that is what it is called, dictates that as they can no longer tell lies and are not yet ready to tell the truth, they had better say nothing.

# The Storwyncyxkycs: an everyday  tale of programme-folk management

There seem to be a thousand and one ways of organizing to manage projects. Very often organizations seem to pass through periods of change as they adjust themselves and try different structures and philosophies. Extremely expensive-sounding titles are applied – 'business process re-engineering' is one such title and I am sure that it is valid and beneficial to many people and many organizations. Some of the people who find it a useful name will be found working for management consultancies who are currently available to help you re-engineer your business processes and your bank balance by charging vast fees, but there are others I am sure. It's just that I don't know any.

| Bitchy! |
| --- |

I have noticed that there seems to be a pattern of growth that begins with an organization with neither project nor programme structure and leads through a number of pitfalls and traps towards Utopia.

Not all companies pass through all stages and no one at all arrives at Utopia but it will help me and, I hope, you if we take a hypothetical company and test out various project- and programme-management organizational structures. We'll take a baby and install various strategies and structures and see how the company grows.

| The number 42 bus goes via Utopia five minutes before you get to the bus stop. |
| --- |

The company will be deliberately forced to make some mistakes in the hope that my hypothetical mistakes eliminate your real ones. It is not a problem to find errors that organizations might make as a short look at a few companies will provide all the cock-ups, errors of judgement and mistakes you could possible need. So as not to embarrass anyone nor shake any skeletons long since dead and cupboarded I'll stick with my little hypothetical errors preventing your real ones. A sort of hypothetical prophylactical.

One other thing – to get across many of the points about organizational structures I am going to use organizational charts. These are sometimes called organograms which can be confused with paper folding and a kind of greeting message not available legally in England.

Such diagrams can only show the formal ideal of an organization and I am very suspicious of them. I saw one on a wall in an aeroplane engine manufacturers design office which showed everyone working on the project with a little picture alongside their box. It was an easy way to find out who the chap near the third window was.

I also have seen studies of communications between people in organizations. These studies examine the frequency with which pairs of people communicate and plot that frequency in some way. One system had all the people drawn around a circle with lines crossing showing who spoke to whom. The more often they spoke, the thicker the line. These diagrams looked nothing at all like organizational charts which really only show the formalized superior–subordinate relationships. Still, they are the best we have.

## Functional structure

Perhaps the worst arrangement is to have no project organization whatsoever. The diagram opposite shows the non-project organization. There is nothing wrong with this company as a functional organizational split is very appropriate, sensible and workable in many situations.

The organization is grouped into a number of specialist departments or divisions. Each department deals with some aspects of the company's business and the departments are co-ordinated by the board of directors or, in a smaller company, the boss.

If there are projects they will have no specific team or leader associated with them and therefore pass from department to department without notice, without planning and without an understanding of priorities. Projects do not sit comfortably within this company as there is no structure to support them. For this reason most things that might be classified as projects are run without the organization. Without means not within, not without the help of. Here is the organization's structure.

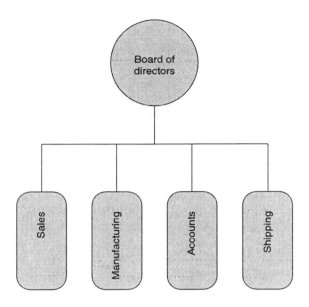

*Management structure: Norfolk & Goode*

Norfolk & Goode are in the toy rifle business. The boss came up with some good ideas for toys years ago and the firm churn out vast numbers of annoying little plastic devices with which small children rush about pretending to be Rambo or Robocop.

The manufacturing team runs the factory which is bigger and more vital than most other departments but nevertheless stuck in between the back of the offices and the railway line. A group of salesmen led by the Senior Marketing Executive (who happens by sheer coincidence to have the same surname as the boss – there are lots of Storwyncyxkycs in Clapham) travel about the country in their Vauxhall Cavaliers, armed with samples of the toy rifle, visiting toyshops. They get sick and tired of being reported by some curtain twitcher in Oldham as this usually results in being chased down the motorway by the police who expect them to be armed desperadoes. Once they escape from the arms of the law they return with orders to head office.

The shipping department receive the toy guns from manufacturing in plastic buckets and sort them into two piles. One pile contains high-quality produce and the other contains rejects where the trigger has fallen off or the colour is wrong. They wrap the rejects in blister packs, pack the blister packs in cartons, pack the cartons in boxes and palletize the boxes before shipping them off to the toy shops. Shipping actually spends more each week than manufacturing. The other pile of guns go back to be melted down but perhaps that is just the way it seems to an inexpert outsider.

The accounts department deal with paperwork. They raise invoices, chase payments, occasionally pay the rent and rates and sort out wages. Whilst most of the other departments are full of men and are decorated with pictures taken from the *Sun* or, noting the lack of clothing, under it, accounts is populated predominantly by women who read *Hello!* and *Cosmopolitan*.

The board consists of Mr and Mrs Storwyncyxkyc sitting in one of the few areas with a carpeted floor. They hire and fire people and meet with the heads of the four operating departments once per month to discuss problems and have a working lunch.

Products are not specifically manufactured to meet orders; manufacturing goes on and whatever is made is sold or trashed. You can see that there is little room within this simple structure for a project. When Mr Storwyncyxkyc decides to build a new factory so that he can step up production, increase productivity and, most vital of all, spend days away from Mrs Storwyncyxkyc he will have a project on his hands.

Despite the fact that he has a great deal of expertise in the company in the shape of people who know all about factories to make plastic toys and account department layouts, he will find that everyone is much too interested in running their own work to worry about a new factory down the road. And why should they get involved? They are on the company production-related target scheme which means that, if they hit certain targets, they get a little extra cash to spend on the *Sun* or *Hello!* as the case may be. Helping out with the new project is just a diversion from what they see as 'the main business'.

The structure does have its strengths. People get the chance to spend a career in one department learning and understanding all the nooks and crannies of manufacturing or shipping or whatever. They can specialize in that area and grow up the hierarchy within that section. As the organization grows and as people leave for any reason, there is usually someone ready to step up into the shoes of whoever has gone.

All the expertise about shipping plastic guns resides in just the one place it is most useful – the plastic gun shipping department. People in accounts are steeped in accounts, they read the fascinating *Accountancy Age* magazine and fiddle about with double entries all day long. They keep up to date with developments in the field like new techniques, computer applications and tax dodges in the Cayman Islands.

They may find life a little dull as the objectives rarely change. The objectives they had last year were pretty much the same as this year and there is a fair chance that next year… Why, by the way, did the accountant cross the road? Because he did it last year. They don't have the cut and thrust of projects to fire up their lives but, then again, what they've never had they never miss. But you try telling that to a fourteen-year-old.

In place of enthusiasm we do get expertise and experience. Some people feel very comfortable in a specialist department and this is especially the case in a very specialist department – a pneumatic section of a machine shop, the paper-handling group in a printer manufacturer and the actuarial departments of insurers spring to mind.

Meanwhile, back at Norfolk & Goode... The new factory down the road is going to be in the wholesale market. When, as a spotted youth, I first heard about this part of the world of commerce I assumed that these were the people who sold holes. It was never clear to me why anyone should buy large quantities of holes from a hole seller. I am older and wiser now and have found this is by no means the most elusive element in commerce.

Mr Storwyncyxkyc has come to the conclusion that his salesmen should call on the larger stores and distributors and collaborate with them to design and develop and then sell them a quantity of some new toy. They could develop the designs and so on and give him an order for 10 000 of the little plastic doohickies. He dreams of making a killing in the breakfast-cereal trade by manufacturing strange plastic beasties with which to choke kids and small dogs – the original cereal killer.

The sales and manufacturing system would be somewhat different – his salesmen would win orders for large quantities of plastic toys and he would get them made up, packaged and shipped. There would be changes to the production machinery for each batch but there would be very few invoices and no more chasing small shops at the other end of the country for small outstanding payments.

As far as the boss is concerned the only difference between the old factory and the new would be the employment of a designer who would knock out the drawings for the production people to make. He plans to have new production tools made up by a subcontractor called Storwyncyxkyc and Storwyncyxkyc plc (no relation).

The organogram for the second factory appears on page 172.

As the new factory building work gets going a benefit of the old functional structure becomes apparent as the specialists in the old place see exciting new jobs becoming available down the road. Noting that Mr Storwyncyxkyc is a bit of a soft touch and noting that no one in their right mind would ever try to touch Mrs Storwyncyxkyc to find out how soft she is, informal applications of many sorts are being made for jobs at the new plant.

Second-in-commands in the four old departments find opportunities to offer their services to work as first-in-commands in the new factory. Of course the designer has to be hired in but the new factory is generally staffed by a mixture of people from the old place reinforced by people from outside the organization.

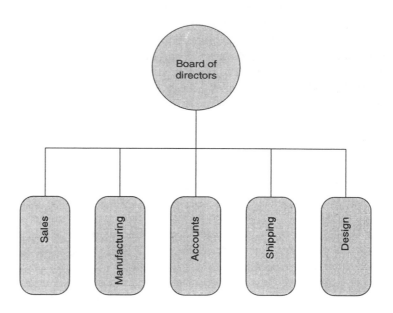

*Second management structure: Norfolk & Goode*

Eventually the plant is nearly ready for production and the salesmen start to bring home the bacon in the shape of enquiries for large quantities of plastic thingies. These are given to the designer who designs tools and the production people work out a price for each job. Eventually the first job is won and an order is placed for 5000 pistols. Things don't change much.

The tools are made up and the manufacturing begins. Shipping are not very busy. In fact not very busy is a sweet and polite term for the shipping people who have as much to do as a statue in the park. They fill their time thinking about the huge packaging assignment coming their way soon. As other orders come in (they hear about these in the canteen) they decide to organize the shipping department to get ready to deal with larger orders. They dismantle the racking in preparation for re-erection in a better location. Just as the last bit of racking is on the floor and everything is ready for the re-erection in the new location, just when the place looks like a warehouse in mid-racking relocation, the 5000 pistols arrive from manufacturing. They have been working overtime to get the rifles finished a week early to make room for order number two which is even bigger and better.

A lesson is learnt – it helps to know what is going on next door. Shipping gets into chaos quickly but is nearly on top of the situation when the next consignment arrives from manufacturing – 150 000

gnomes for a muesli manufacturer – which throws them back into a state of panic.

The inevitable happens and some gnomes get sent to the toy shop where they sell quite well, some pistols go to the muesli factory and others get lost altogether. Chaos rules at Norfolk & Goode. Things are Norfolk & Goode at all.

Invoices arrive at the client's offices long before or long after the goods but are often returned as they contain enough inaccuracies to make the weather forecasts look spot on. The dissatisfied clients call the only people they know at the plant – the salesman and Mr Storwyncyxkyc, who is already tearing out his eyebrows.

> He tore out his hair years ago.

Priorities get switched daily and sometimes hourly. Only occasionally are priorities changed consciously. What has happened is that they have a purely functional structure trying to run projects. Projects fall from department to department without planning and without notice so every time a project arrives in a department it comes as a surprise. Projects often get lost in the melee all together and priorities shift like snow in a blizzard.

No one can plan their workload as they don't know what to expect. Sometimes an urgent job arrives just as the annual outing to The Tooting Tap and Die Club gets underway. Some days everyone stands around ready and willing with nothing to do.

Motivation amongst the workforce is low. Respect for the management is lower. Every packer, shipper, gun moulder and gnome finisher has the greatest excuse to go on reading the *Daily Mail* for another half hour: there is such chaos that no one will notice or care if they take a teabreak, break their necks or break out in a sweat. Why bother?

There is no horizontal force or co-ordination across the department looking after the projects. Actually there are three groups of people trying to make up for the lack of a project-management role. The salesperson's commission depends on getting the goods to the clients and the clients know the sales team – it was the sales team that demonstrated the gnomes in the first place and landed the order.

So when the client wants some action like an answer to a simple question such as: 'When will our dinosaurs arrive?' they call the saleswoman. She searches around the office and factory to discover the state of the project. It is eventually found as a bunch of drawings forming a design scheme behind a radiator in the toolmakers corner. No one knows how it got there or how long it has been there but a family of previously unknown small insects have moved in. The saleswoman learns that no one knows what priority was attached to the job.

As each client calls in to complain, his job instantly becomes the priority project, something else gets dropped and, for a short time, the new project is worked upon. This waste of time is stunning as the designer and toolmakers are often picking up a job they last saw three weeks ago, devoting time to learning about it once again, finding the incomplete drawings and tools and generally getting back into the task. The delays caused by 'getting up to speed' are repeated time and again as jobs are picked up, dropped and picked up again.

Respect for the management is at an all-time high.

> As some Americans don't understand irony, this means everyone hates the boss.

In fact everyone blames everyone else for the difficulties. Everyone knows that things are not going well but no one understands what can be done. Wars break out between departments as they hurl complaints and accusations at each other. The shipping department suffer worse than any other group.

Shipping are the classic 'last-in-line' group. As projects get delayed in departments like design, tooling, manufacturing and so on and in between these departments, the amount of time remaining between today and the fixed and unmoveable delivery date gets shorter and shorter.

The delivery stays fixed as today moves inexorably forward which would be fine if a day's progress was made on each day. The days pass whilst very little happens to the project. Eventually the project is passed over to the shipping department who actually get the goods out of the door onto a lorry and to the client. Because they actually deliver, they are seen as the ones who are late and they are late, but this is a case of shooting the messenger.

Packaging and shipping becomes a slick and smooth operation able to deal with almost anything thrown at them, they have sophisticated equipment and slightly too many people for the 'normal' workload plus a group of itinerant workers they can call on when needed. Nevertheless, as delays occur up the line the weight of responsibility falls on the poor old last-in-line.

Only three things can happen to make things worse and those things all involve a computer. If the boss rushes out and buys a heavy-weight local-area-network based project-management system at this point to mend the situation I think we can safely assume that the business is going to go belly up. He could hire a management consultancy firm to examine the company but the fees alone will kill all possible profits. The third thing? A management consultant gets hired and recommends the purchase of a heavyweight local-area-network based project-management system. Ye gods and little fishes.

*Norfolk & Goode*

Saved by the bell the boss notices something. On one project, a new salesman followed it through the organization letting people in the various departments know what was happening in advance, what to expect. This one job went well. The salesman soon got sucked into the whirlpool that is Norfolk & Goode but the one successful job left an impression. Even the head of packing and shipping was impressed.

These orders are actually small projects and the company is making a number of elementary mistakes. It has no one interested in pulling the project across the departments. Every one is interested in keeping their department going as efficiently as possible and they have become quite used to dealing with the daily chaos that gives them so many excuses for anything and everything that they have no real worries. Life is frustrating but easy. More time is invested in the creation of excuses than of gnomes.

The company decide to go for a matrix. Matrix management has earned itself a bit of a bad name but actually most people do use something rather like a matrix. Please don't get put off by the name, look at the concepts.

## Matrix management: a beginning

In this environment a hierarchy is created dealing with the specific projects and which sits alongside the normal functional structure. Mr Storwyncyxkyc finds a diagram that typifies a matrix-management organization and this really appeals to him.

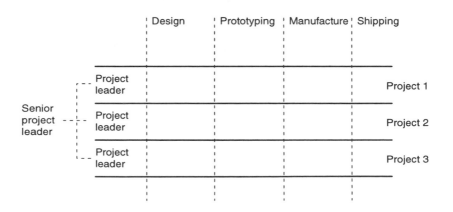

*Matrix-management structure*

Across the top of the matrix are the various functional departments and down the side are the various projects. The sketch is over the top for Norfolk & Goode as there is a two-level hierarchy of project people which would be far too much for his firm but the basic idea seems sound.

He takes one of the salespeople, Edward, the one who showed some ability to get a job pushed through the organization but is not that great at selling, and promotes him with the grand title of project manager. He announces this at one of his working lunches to the consternation of his five departmental managers. They all have some difficulty seeing what this salesman is going to do in his new role.

Edward takes to this new role well. He researches the topic and realizes that his role is to plan ahead and get other people to do the same. He buys a copy of *Supertimeworkprojectschedule* Version 9.2 for Windows & Doors and goes on a three-day training course learning how to use it. Somewhat belatedly he goes on a three-day course on the basics of project management. For this there are two reasons.

Firstly, whilst on the first course he finds out which buttons to press to achieve something but learns little about why that something is worth achieving. The second course clarifies what he can sensibly achieve. Edward realizes it would have been much better to attend the two courses in the other order. The other reason is that on the first course he and a fellow student called Pippa spend a great deal of their free time happily together and both agree that the only sensible thing is to go on a more general course together where, once again, they spend much time together. Isn't life a beach?

Back at the factory, Edward, separated by some 250 miles from his new friend, gets down to work. He draws up a plan for a typical project which has five tasks – one to cover the work in each department. This he considers too simple. Not too simple to control the job, not too simple to communicate what is going on, too simple to justify his role and salary. He therefore calls his five tasks phases and breaks down the five phases into tiny detail. He knows something about the process involved as he studied it closely so, sitting in his little office tapping away with his computer, smiling at comments about playing computer games all day, he develops a plan with around seventy tasks.

He gets the five departmental managers together and explains what he hopes to achieve and what he has done. He explains that this is a typical plan for a typical project and that he would like to use it on a test project. Sure enough the next project is awarded to the firm partly because of this detailed planning. The new client's reluctance to place the order with Norfolk & Goode is based on the company's appalling record of late delivery. This is overcome by the salespeople when they talk of the extremely sophisticated project-management system installed in the company which will almost certainly guarantee delivery on time.

Edward's typical plan is called into play, the durations are modified to suit the real project and the resulting plan is used to impress the client's socks off.

This plan is issued to the five departmental managers who at first ignore it completely. They ignore it at first because they don't understand what it says, they were not involved in producing it and they don't really see how it can help. Edward runs ahead of the project telling people how the project is going and when they can expect to get on with their bit. He monitors the process through each department and notes down variations from plan. At first the model project goes very well and the general view is that there might actually be something in this project planning. Regrettably a big client calls demanding his plastic footballs to coincide with the FA cup advertising campaign and Edward's well-planned project's priority drops through the floor.

Edward tries to explain all this to the boss when the test project is finally shipped six weeks late saying that the project was going well until the planning was over-ruled and that he can't be expected to make projects go to time when the boss simply goes over his head. The boss thought that one of the benefits of being the boss was the ability to go over people's head whenever he felt like it. A number of problems have arisen, let's look at them:

Whilst Edward has no authority over the departments he does have responsibility for delivering the project. This is an invidious position and one you should avoid if you value your career. His role is to deliver the project on time to the client and yet he has no resources under his command with which to achieve the objective. He has to borrow and steal, beg and plead to get his work done. He has no chance when over-ruled by a higher authority.

The boss expected too much. When he angrily said 'We brought you a computer, we sent you on those courses and it hasn't helped a bit', he was expressing the belief that buying project-management software and hardware can solve problems. Actually it does the opposite – it raises problems. Edward considers this response: 'At least we know that we are late and we can tell the client in good time or do something about it.' Edward is right, planning only helps to raise problems in good time so that something can be done. It does not get problems out of the way. Edward is very likely to get sacked for saying something which the boss wishes so strongly not to hear so he keeps silent.

Because the new project was a pilot of the new approach, everyone elevated it to a higher status than it would otherwise deserve. For this reason it went very well – simply because everyone knew that it was under scrutiny. It nearly worked and would have if the boss had not kicked the project into touch by giving the plastic footballs even higher priority.

The plan was too detailed. As soon as the plan collapsed, department heads came out of the closets in which they had been hiding waiting to see how things panned out complaining that the plan 'told them how to do their jobs'. Edward's plan dwelt in some depth on the detailed processes within each department mainly because he wanted his plan to be impressive.

Also the plan 'belonged' to Edward. Everyone else can conveniently and rationally blame the plan as they didn't put it together, Edward did. They don't feel it is *their* plan, they feel it is *his* plan. And he, they note, keeps leaving early on Fridays to see the girl he met on that training course.

The plan ignores the availability of resources to perform these tasks. The plan simply lists tasks and dates so that people can know when a task is due but not which of the designers, machinists or packers was supposed to do it.

Edward's future is looking as bright as a coal miner's and he will need some help to overcome these problems. He manages to convince the boss that even more detail is needed and that he must get into resourcing. Cancelling a weekend away in Leeds he gets into resource allocation, aggregation and smoothing. Eventually his plan is refined in that:

1. It has even more tasks.
2. It specifies who does what by name.

On Monday Edward proudly shows this to the assembled functional managers who reject it out of hand. They didn't like the first plan as it seemed to be unnecessary and encroached on their areas of responsibility but this one assumes too much by far. This plan tells them which of their workers is supposed to be working on this one project. The head of shipping gets a good laugh at Edward's expense by telling him that he can't put Fred on this job at the time shown on the barchart because Fred will be away on holiday with his kids. Edward's standing sinks still lower. It is now knee high to a centipede in a muddy field.

Edward has gone too far in his efforts to help and has stamped on the areas of authority that rightly belong to the department heads. Later that week he goes off to be consoled by Pippa in Leeds.

He returns next Monday a new man with a new idea. He tells each functional manager that he can see how he got it all wrong and asks for their help in getting it right. Can they muck in and decide on a strategy and some objectives and give planning one last chance?

They decide that planning can be useful if they all plan together. If they all plan together they will at least know who is doing what and what they will be expected to do in the near future. Best of all they will be able to argue with the boss when he wants to switch priorities.

They go back to Edward's simple plan with only five tasks and refine it to fit in with their own thoughts. Then Edward gets a list of all the current jobs and builds a multi-project plan showing one plan for all the current work – twenty different projects – each with ten tasks. The projects are offset by their respective starting dates and they allocate skills to each task. Skills, Edward explains are trades like packers and machinists, not individuals like Fred and Sophie. They use the past bonus sheets to calculate how many machine-hours and packer-hours are likely to be needed for each task on each project.

Edward calls Pippa to tell her that the plan they hatched over the weekend seems to be working. Pippa wishes that Edward would talk about something other than project management and plastic rifles.

The new, simple plan for all the projects with their skill requirements produces a series of histograms which prove conclusively that the workload cannot be achieved with the current resources. This is no surprise to the departmental heads but it does give them some ammunition.

The whole group faces the MD with this news and explains the information on the table. He appears greatly worried but is secretly delighted. He acts worried as he can see that he is going to have to let down some customers. As someone points out at great personal risk, letting down some customers is new – previously they have been letting down all customers. The boss is privately very pleased that they are all planning together and this has to be a good thing.

Edward's plan looks very simple but is widely accepted and respected. Everyone feels that they had a part in producing it and that they in part 'own it'. They even defended the plan to the boss. Edward is no longer planning in his little office, planning gets done as a group and Edward takes the ideas and types them in and generates the results. They agree on a strategy which involves letting down some rather less important clients, meeting delivery dates on the bigger orders and also involves hiring in some design staff in October. The staff agency cannot believe their ears when they are asked for some designers in two months' time when previously two hours would have been a long period of notice.

Some of the clients that are to be let down are told there is likely to be a problem and some appreciate it. One cancels the order and goes elsewhere but they all agree they would have gone elsewhere next time anyway.

The salespeople still make ludicrous promises to potential clients who are not as well planned as Norfolk & Goode so when the next new job comes in the boss drops it on the designer's table with a long tale about this new client's importance and how it could mean loads of extra work and the designer points to his barchart and asks: "OK. Which other job shall I drop?" Later at a planning meeting, they do decide which project to drop. They drop the new one before picking it up.

I was tempted to tell you that at the planning meeting they found a way to absorb this new project because things were working so much more smoothly that they actually had some spare resources but I figured you wouldn't accept such a huge swing in attitude that quickly. Life is not that easy.

After a few weeks of these halcyon days Edward notices that the plans are disappearing from the walls and chaos is beginning to rear its ugly head again. The problem is well summarized by the head of manufacturing: 'It was great but it doesn't show the new Supersonic Grenade job, it still shows the Crunchy Berryfruit Motorcars freebie and we already did half the work shown on there. It's just out of date'.

Edward realizes that he has to keep the plan up to date so he sets up a system where each month, the five heads of department sit around a table with an update of actual progress within the previous month. As they talk through each project he taps away at the computer coming up with different priorities and schedules until they eventually agree on the next month's workload. They also run through a great deal of practical stuff at these planning meetings because they are in one room thinking about the future.

Time passes. More time passes. It is a few months later and everyone is accustomed to Edward handing out the barcharts for that month and collecting the old ones back in. He is a little late with the issue of that month's plan as he has been away on his honeymoon with the girl he met on the project-management training course, which at least means he won't be zipping up to Leeds every Friday night. She finally got him to talk about something other than project management.

Everyone has a small barchart on the wall showing their workload – it shows the jobs they are responsible for in all projects. He prints, each on one A4 sheet, the following:

- all design tasks;
- all tooling tasks;
- all manufacture tasks;
- all packing and shipping tasks;
- all delivery dates – this goes to the MD and all salespeople.

These get issued every month and quickly people come to rely on these reports. When someone asks when the toy fax machines will be ready to be packed, the manufacturing supervisor consults his barchart and replies positively and with confidence. Projects still get delayed but they know about these delays and deal with them as efficiently as possible. They try hard to work together and co-operate.

When people talk about something being 'ready to be packed' they all mean the same thing. The seemingly innocuous phrase 'ready to be packed' has meant different things to different people for some time

and the process of getting planned has created a common understanding of the stages in the project life cycle. When clients call in for
status reports on their projects they get consistent responses – these
are not necessarily honest but they are consistent. The matrix is
working.

## Matrix management: the next stage

Norfolk & Goode, helped by their workable programme-management
system, look after all the projects in hand. The departmental managers
regard Edward as a useful service department but he is being worked off
his feet. His wife, who has now moved down from the North has to
watch TV all evening whilst Edward either works or snoozes on the
couch. The company decides to grow the matrix.

Edward is promoted to programme manager and he recruits three
project managers. Each project manager takes responsibility for roughly
a third of the projects but once again the division of responsibility issue
rears up – are these project managers going to have responsibility but no
authority or is there some other way?

His boss has a real thing about management consultants. He says that
he once hired one to answer some questions and all he got was more
questions. This is exactly right, the client who recruits a management
consultant has a problem which he hopes can be solved. The management consultant realizes that if he actually solves the problems
(assuming for a rash moment he is competent to do so) his assignment
is at an end. The head of the consultancy rewards people on work they
bring in and what easier way to bring in more work than to study the
client's problem in great detail, define the problem, classify it and submit a report containing seven further issues which require long and
expensive investigations?

Someone said a management consultant was someone who borrowed
your watch, told you what time it was and then charged you for the
privilege. Despite all this they do recruit a management consultant to
study their matrix-management problem – the question succinctly put
is: how do we proceed? The consultant explains a number of choices.
She gave them a short lecture on the different forms of matrices in use.
This is what she said:

There are a number of workable practical solutions and an even larger
number of completely impractical and unworkable solutions. We'll tend
to concentrate on the former but be warned, what is practical to one
company may fall through the floor in another.

She talked about four major roles that exist within a company:

**Table 3.1**

| | |
|---|---|
| **Project manager** | plans and manages but doesn't do anything directly |
| **Functional manager** | runs a department full of bright young things experienced in a specific area: design department, testing lab, installation group |
| **Programme manager** | manages the project managers |
| **Resource or operative** | those that actually lay bricks, weld pipes and write code |

Using these four roles she went on to describe various matrix variations. She was a bit of a role modeller.

I think we should leave poor old Norfolk & Goode, Edward, Mr and Mrs Storwyncyxkyc for the moment. I have tried to paint a picture of the mistakes and pitfalls that can and do often happen to an organization getting into programme management so these players have served their purpose.

It is time to get back to theory and discuss some matrix-management structures in a little more detail and examine the issues the management consultant would examine, but without the fees.

# Matrix-management formulae

## Types of matrices

Let us first of all take a look at the organizational matrix in its classic forms. I'll briefly run through the ones with which I am familiar, trying to state the advantages and disadvantages as I go.

### Subcontract matrix

In this arrangement you give the project managers a budget and let them 'buy' work from the functional departments. Each project manager is given work to do, in the shape of projects, by the programme manager. Work is farmed out depending on the availability of the project manager's time, knowledge of the client, knowledge of the type of work, conflicting holidays and other workload.

Often the project manager is dragged along to the client's office to be shown off. 'Fred here will be the project manager allocated almost full time to your job, Mr Big Cheese, and he has been handling projects just like yours for ages.' In this statement 'almost' means 'around 5% of his time' and 'ages' equals 'since late this morning'. That's salesmanship.

In a subcontract matrix once the project manager has got the job he works out a budget and a project plan. These two documents may be quite simple and he may only need to plan down to the first level – perhaps only five or ten budget items and tasks. The project manager then hawks his plan and budget around the various internal functional departments asking 'Would you like to do the design for this job?' or 'How about the moulding of this new gizmo?'

Informally, as he is within his own company, he is seeking quotations for executing the various stage of the work. The functional departments essentially bid for the work. As with subcontracts, the project manager does not get involved with who actually does the work and may meet no resources face to face. He may not know when the actual work is to be done but as long as it fits within his project milestones he will be happy. The functional department manager takes away the job and comes back some time later with his part of the job done. The project manager accepts the project back and passes it along to the next department.

Where a project and a functional department meet on the matrix is sometimes called a **cost centre** or a **work package**. A cost centre refers to a package of work being performed by one department on one project.

This approach works exceptionally well where there is more than one functional department that can compete for the work. Small working groups are encouraged to set up shop within the company and punt around for work. Within the same organization money rarely changes hands. The project manager has a budget, part of which gets transferred to the functional departments. The functional department manager has a profit target so the artificial 'income' from the project managers gets balanced against the only too real wage bills and invoices for hardware to calculate his profitability.

Motivation can be a bit of a worry under a subcontract matrix. The departmental heads are motivated towards making a profit just as if they were external contractors. Perhaps the interests of 'quick and profitable' for the functional department and high quality for the project manager are going to conflict. Also the functional managers are going to prioritize those jobs that they see as helping their department and its budget along. Such prioritization may not been in the best interests of the company.

The project manager tends to be a little distant from the work in this arrangement. Even in-house, the project manager may never get to meet with and discuss the problems with the operatives actually working on her project. The project manager need have no right to get involved with resource allocation at all. The job might be done by everyone in the department in one day or by the tea boy over the next three months but as long as the agreed milestones are met, the project manager has to be happy. If the functional manager permits, the project manager may meet with the people working on her assignment.

Another little disadvantage of this approach is that there are no highly-motivated, enthusiastic project teams – the project team does not exist. However, on the plus side, the functional departments get very good at carrying out project work which arrives and leaves rapidly and frequently. They work within their own specialist field and are surrounded by like-minded people with whom they can beg favours and give advice. The specialist departments become centres of excellence within which expertise on a specific topic is developed and maintained. People spend their lives climbing the functional ladder becoming ever more senior in their roles. Junior assistant gnome designers become assistant gnome designers and then later senior gnome designers before rising to the dizzy heights of head of the gnome design department.

In some organizations the project manager can discuss her needs with contractors outside the company. The internal department is in competition with a number of outside agencies who would be only too happy to take on some of the workload. Indeed some of the better people within the company leave and set up their own small, efficient companies to tender for work. Instead of a standard and probably friendly agreement between the project manager and the functional manager one will need a formal contract, but the principle holds firm.

What happens if a department is held up and causes delays? As long as those delays stay within the agreed milestone plan there is no problem to the project. But, life being what it is, the delays will probably mean that some resources are engaged for a longer period than expected and therefore they will not be free for their next job. As soon as a functional department realizes it is going to miss a deadline it should tell the project manager, who will tell the other departments down the line about the delays.

The project manager has every reason to be honest. Often the functional departments don't realize or won't admit that they are going to be late until it too late to correct the situation and then the next-in-line gets upset at being dumped on from a great height. The person doing the dumping is the poor old project manager who has to go to the next department and tell them that as the design for the new product is not ready, the test assembly work cannot begin on schedule. The project manager got dumped on in the first place by the functional department who screwed up the design, which makes the project manager feel warm and confident about the design manager.

For non English speakers that was an example of sarcasm.

In a competitive world you can guess which design manager is going to the bottom of this project manager's list for the next job. In a non-competitive world, bottom of the list equals top of the list as there is only one name on the list.

Let's take a look from the departmental manager's perspective. These people are sometimes known as functional managers or resource managers. Departmental managers can help by planning their work. From their perspective they have many project managers who bring them work and they have to try to satisfy them all. Each project manager, each job, starts as just two milestones: 'start' and 'finish'. Start means the start of this department's role in this project and finish means the end of this department's contribution.

The department can take a list of such simple dates and get busy planning their own workload. The departmental manager looks at the sort of work involved and, knowing the people under his control, decides who is going to do what and when. He assigns Sue to the new Football Simulator project, puts Joe on the new teapot concept and gets Alan to do the chairman's freebie give-away nodding dog for cars' rear parcel shelves.

The departmental manager does this kind of thing balancing the operative's strengths and weaknesses against the urgency of the job, their holidays and training plans and the non-project work he must never forget. His plan starts off with a list of unconnected jobs, gets broken down into small subprojects and then gets extended with resource allocations.

At this level the functional department might be down to work planning or work scheduling and a simple Sasco, card index or Lego planning system might work well.

## Full-time assignment matrix

This is another approach and a much more personal one at that. The project manager is once again given the job to do and approaches the functional managers to borrow staff. Conversations start like this: 'I'm going to run the dam project in Malaysia and I need a concrete technologist, 40 carpenters and some good luck.'

The functional manager thinks about this and gives the project manager a concrete technologist person for the duration. How the roles are changed! Instead of doing the job for the project manager like a sub-contractor, the functional department is lending out bodies like a bodyshop (otherwise known as a staff agency and not the wonderful-smelling shops that sell almond and strawberry foot massage oils).

The project manager is building a team of people on loan to him from the various specialist departments and the team will set about this project as one united group. The project manager must clearly plan in detail enough to predict his demand for resources of all kinds. His budget will be hard hit if he has carpenters and concrete technologists sitting about on his dam project waiting for some dam thing to do.

We are typically talking about quite large projects where the team are working full time for the duration of their time on the projects. Very often there is a removal from the head office to a project office. The project manager builds his team and tries his best to weld them together to work together and achieve the project. This is what happens every day in construction and heavy engineering all over the world. The functional manager's job is to have the right sort of people just about to come free when they are needed.

The ideal departmental manager should run a tight department where people move from project to project with a minimum of breaks. She would provide advice to her departmental staff members about their careers, their training and what will be good for them to do. She hires in new people, bids a fond farewell to good people that leave and breathes a sigh of relief when not-so-good ones leave. Whilst there is a pool of expertise in such organizational structures, it tends to be spread around the place, country or even world.

Departmental managers will therefore organize 'knowledge exchanges' bringing together all the concrete technologists from the 14 projects in hand around the globe. At these events the company's total experience in a topic is collected in one room so they can swap ideas and experiences. Our man in Malaysia might give a lecture to his fellow specialists about the special problems of laying concrete in tropical conditions. Perhaps someone else is working in the office on a different but equally interesting technical problem.

She keeps in touch with the many project managers and especially when one of her specialists is coming to the end of a job. She must find something to convert the unemployed resource into an employed resource – ideally another job. Her role is to balance having people ready to drop onto new projects with a low running cost.

The project manager is rather well off – he has a full-time team who are likely to become quite excited about the project as they live, eat and breathe it every day. Not for them the diverting life working on 42 projects with as many managers. If you want to enjoy your work life and to become deeply involved in it, there is little like a team resident on a major project to get the juices flowing.

A slight problem is that every resource has two bosses, two lines of authority. Somehow the project manager and functional manager must manage the resources between them. Clearly, the long-term future and career path of an individual should lie with the functional manager as she has a long-term perspective on the person's future. The project manager can say, at any time, 'Sorry Fred, but this one is going to be built in steel after all', and pass his concrete expert back to the functional manager.

Equally, what happens if the resource is a poor timekeeper or fails to perform in some other way? Who issues the warnings and eventually plays the card game where a UB40 is a winning hand? Such issues need addressing in this two-boss environment.

## Part-time assignment matrix

A much more common situation is an assignment matrix as described above except that people are not loaned out full time for the duration of the project to a project manager working on a site in a distant land. As you must have guessed from the title, people are loaned part time to many projects.

In such an environment you would normally expect to find quite a few projects in hand in the one organization at the same time. Once again specialists are kept together to share their expertise and so as to provide a career path in their chosen specialization. Once again the project manager exists to look after jobs that are being undertaken against a background of, most commonly, other projects and a whole raft of non-project work.

I visited the headquarters of a major high-street store that was structured in a part-time matrix. Functionally they had a number of departments whose normal workload kept them busy enough thank you. The functional department list included:

| | |
|---|---|
| **purchasing:** | buying in the goods |
| **distribution:** | the lorry fleet |
| **retail:** | running the shops |
| **IT:** | central computing |
| **accounts:** | who knows what they do? |
| **wages:** | making out the pay cheques |
| **architects and building:** | acquiring and building shops and warehouses |
| **personnel:** | keeping everything staffed |
| **training:** | keeping people educated |
| **directorate:** | same as accounts |

There were a number of projects in hand when I visited and one project was a new North of England warehouse. The warehouse was not typical but was, potentially, the worst project in the history of the company so I'll use it as a juicy example.

The warehouse project involved most departments in the company. Distribution were going to use it as a Northern lorry park, workshop and garage. The architects would find the land and build the shell and fill the building with one of those automatic racking and delivery gadgets. If this system delivered the goods to the loading bays at the correct height the vans would take the stuff to the shops,

IT were going to extend the nationwide stock-control system and personnel were going to recruit a warehouse team to back up the senior manager, who was going to move home to live there. Wages and accounts would need to modify their systems to handle the new location and training of the new staff would probably be necessary. A project manager was appointed for the warehouse project for which we should be duly grateful. He contacted the head of each department to ask for a contact name, the name of the person who would be dealing with this major new project.

Answers ranged from the global 'No idea, do I need to decide?' to the more interrogative 'What warehouse?' Eventually the poor old project manager assembled a team for this project. He visited each departmental head in turn and explained how this was to be a high-profile project which should make a real difference to the way the organization works.

He explained his need for input from the department and asked for a bright intelligent person to spend a small part of each week on the project. The project manager explains how this was a chance for someone to shine within the organization as 'the new project will make waves throughout the group'. The functional head, ever conscious of his budgets and targets, thought about all this and suggested the name of the person he could most afford to do without. As usual the project manager asked for the best person and got the worst. After negotiation a sensible team got assembled.

The last couple of projects had been such an unmitigated failure that everyone suggested someone who could actually help. The proud project manager printed out his project team and sent this to everyone including the distant MD. The project manager's aim was to make the team feel special, important and part of something good and wholesome – he was trying to build a team.

He organized a first project team meeting. Once again he wanted to make the team feel important so he organized the boardroom for the meeting and laid on prawn sandwiches, soft drinks and a little white wine. Only half the team showed up as they had those conflicting demands from their normal workload to deal with. Those that did show up seemed interested and motivated as he walked them through the project but they were regularly interrupted by phone calls, messages and urgent whispered conversations about issues having nothing to do with the project and possibly nothing to do with the running of the company at all.

Our trusty project manager ended up rather disappointed with his first effort as he did not feel he had really welded the team together into a co-ordinated fighting force. His family were rather less disappointed as he brought home all the prawn sandwiches, soft drinks and white wine left over from his partly-attended meeting.

His first venture into a part-time matrix was not looking good. The net result was a part-motivated team and a well-fed but inebriated family. He went on to try other approaches. He arranged meetings in a local hotel to reduce interruptions and to stress the importance of the event and this worked to some extent.

In this retail-store case there were few projects and each project was a little special. There was a problem to do with objectives – the functional managers needed to hit their targets each month or else face the witticisms of their seniors at the next board meeting. Explaining that 'I had to lend my best two men out for two days a week for the whole month' is seen as an excuse.

It is not hard for project managers to run around trying to get something to happen, anything to happen, until they finally run themselves into a frustrating hole in the ground. The project managers are keen to succeed, they have been given a chance to shine and yet they cannot inspire enthusiasm amongst the so-called team they have been loaned.

Hey, I'm giving this part-time secondment matrix a real pounding, it must be possible to make it work. If you are the project manager trying to make a one-off project work within an otherwise functional company, you must seek senior management support for the project, for this makes a real difference. When the functional managers are told to help with the project and to release people to take part in it, they do.

If there is one message that has come across more strongly than any other in my discussions with organizations bringing in a programme-management approach or running a project in an otherwise functional organization it has been this: get the support of your senior management. Only they can break the mould and let you have the people that can make the project work.

The lack of senior support, the very part-time and unenthusiastic support of the various functional departments is entirely common and could be made worse if there were a few other projects going on simultaneously.

The warehouse did get built reasonably on time and to budget. It was helped by the lack of set objectives so the project manager set his own targets after some time, included some fat to give himself some elbow room and mixed his metaphors with ease.

## *Matrix-inspired issues*

Before leaving the matrix topic there are a few other worries and concerns I want to share with you. They are mentioned here to help you spot trouble rising like a prehistoric monster out of the bog that is your organization and also to help you see how you can manage your own climb up the slippery career ladder.

## *The project-management trap*

The project manager in the retail firm with the part-time secondment matrix nearly drowned in the project-management trap. The trap is laid by getting some poor schmuck to take responsibility for achieving an objective like a new warehouse or a new product launch. The trap is set to go by ensuring that the schmuck has no authority over the only resources available to perform the work.

The warehouse project manager was a classic – he had a clear objective and he had no choice but to use the expertise in the other departments but he had zero authority over them. In a local council I visited, the workload was almost entirely project orientated. Everyone in the group, and this was a group of nearly one hundred people, specialized in some area of council work and was involved in a number of projects. Anyone could be a project manager.

So if you talked to a highways engineer he might be a project manager on one job, an assistant project manager on another, a highways-design resource on three other jobs and have a small administrative workload as well.

I can see that this would work in some environments. As everyone is involved in projects everyone must help each other. If the highways engineer, working as a highways expert, lets down an architect he can expect retribution. If the architect, working as a resource, does well with a design, he can expect help from the project manager when the tables are reversed.

The danger is that everyone gets pulled down to their lowest common denominator – they all do really badly and keep quiet about it so that little gets done, everyone has a quiet life and the council funds continue to be wasted in much the way they have always been wasted. Hoh, hum. It is very hard to sensibly prioritize work in this environment. The oldest and most popular form of prioritization is adopted and that is:

'The project manager that shouts loudest, gets.'

In such an environment the project manager with the most strident voice (or perhaps the most persuasive) will get their job done, often at the expense of other projects being run by less eloquent people. This is fine and dandy as long as you recognize that this is the system and as long as the high-priority projects get allocated to the vocally-enabled project managers.

Pointing out this state of affairs must be done carefully but can be beneficial as few will have recognized the way the organization is run. It is particularly helpful to explain this to the more quietly-spoken members.

## *Non-project work*

In these matrix environments most departments have a significant non-project workload which is easily overlooked in the planning phase. Whilst most people are paid 40 hours per week they have this annoying habit of wasting quite a lot of the time doing useless and non-productive things like eating, going to the lavatory, going on holiday and off to training courses and, worst of all, filling out timesheets.

If you plan to use 100% of anyone's time you are likely to be in trouble from day one. There are two approaches to dealing with this:

1. Build your plans on the basis of project time. Plan your resources on the assumption that they will actually be available to do project work only part of the time. You might be able to look back through timesheets and other data to find out what this actually has been in the history of the company – 80% of paid time would be high, 70% would be normal.

2. Build your plans on the basis of paid time but plan the non-project work. If it is too embarrassing to admit to yourself, your boss and your colleagues how much time is actually not put into projects, you can plan the non-project work. This is especially useful if the non-project workload is not consistent. Training courses and holidays can be regarded as tasks which absorb the resource full time and are fixed requirements. If one testing engineer does a monthly test of some equipment for the first three days of each month, this can be shown as a task. You could have a continuous, everlasting task absorbing 20% of each person to cover the non-project time.

I've been careful to refer to 'non-project time' as this is less contentious than 'wasted time' or 'non-productive time'. They are such emotive terms.

## *Priorities*

Clearly the many project managers and departmental managers are likely to get into conflict over priority. Each project manager will be a highly-motivated person doing their level best to get their project out of the door on time, to spec and to budget. This often means that functional departments get overloaded and cannot cope with the demands of the many project managers.

The first hope is that this comes to light with a little notice. A phrase like 'Look, we just aren't going to able to cope in July' is much better heard during May than on the 30 June. That's one of the things planning is for.

The second hope is that there is some refereeing or umpiring system. There should be some system for deciding on matters of priority when the inevitable clashes occur. Where do we find a referee without short black trousers and a whistle?

The departmental or functional manager has the problem – she has loads of tasks for loads of project managers and can prove that she has insufficient resources to do the work requested of her. Correction, she can prove that she will have insufficient resources to do the work – we are talking future tense here. But she does not have an overview of the other departmental workloads and the political issues. If she delays one job how does that affect the workload of others? Which client is going to be disappointed and how much are the lateness penalties. Should she hire in some help?

The project manager is not to be trusted at all. She is singlemindedly working towards getting her project finished on time and may beat the table to get her way. What we need is an independent body of people who are able to settle matters of dispute. Such a body should understand the workload in terms of projects and the workload of each department. It must be familiar with interactions between projects and departments, projects and projects and departments and departments.

Clients cannot be overlooked in this equation. So we have the programme director, the programme board or some other authoritative body with a grand title to make decisions about priorities. The programme board might contain most project managers and departmental heads plus some senior managers to manage the programme.

## Which skills?

In road projects do you take a highways engineer and train him in project-management skills or do you take a project manager and add, or accept the lack of, highway engineering skills?

The transfer of project-management skills from industry to industry is accepted by most. Telecommunications companies especially seem happy to recruit builders and civil engineers to help with project-management issues. These importees know how to use a phone but little of fibre optics and satellite dishes. Telecoms is forced into this situation as there are few project-management people versed in that industry. It is a one-way migration. Very few builders would hire someone without a building qualification but I know a few ex-builders who have swapped their dirty portakabins and long hours on big projects for a more sedate and comfortable, better-paid life on relatively tiny projects.

I discussed this issue with Dave Kerr and Mike Stubbs – project-management people working on railway-signalling projects with Interlogic Ltd. Dave asserts that 'Yes, we can import some project-management skills but there are some types of projects that demand someone who does understand the nature of the work'. He remembers that 'We started off with engineers taken out of their normal roles to become project managers but they were more involved with the product than the process. We therefore created a separate project-management group out of the engineers, made up of people who are interested in the project issues themselves.'

Mike Stubbs recalled the influx of project-management people:

We were very good at technical issues so needed no more signalling engineers but we were very weak on project management. There were other organizations downsizing, BT was one, thereby putting project managers on the market some of whom we recruited for their project-management expertise. We took on 10 or 12 people like this which allowed us to stop using BR signalling engineers on non-technical issues and replace them with new people. The department was under pressure to grow at this time. The top project-management people tended to be BR people with a project-management interest supported by project-management people from outside.

Signalling is a life-time job simply because it is so safety orientated. It is a career in its own right – how could we bring people in from the outside to become signallers?

If you have a pool of people all working on lots of projects I think a central planning function would be fairly important. The planning office would serve all projects and prepare plans working with the project managers. If these were resourced showing people by name working as projects people and as resources, you get some idea of who is going to overloaded and who will have free time.

We'll talk about project offices in the next section.

## Other matrix terminology

There are other variations on the matrix-management theme which you might like to hear about. Some are different terminology for the same ideas but, to avoid you looking more stupid than usual, it will be sensible to introduce as many terms as possible.

### Lightweight, balanced and heavyweight matrices

These three terms are used to describe the balance in the strength of authority as it rests between the project manager and the resource or functional managers.

In a lightweight matrix the project manager is weak in terms of authority. Perhaps the organization selected a young, enthusiastic but not senior person as project manager. Perhaps it gave the project manager insufficient authority or backing. In the worst case both would be true. The project manager is in that invidious position of having no authority but total responsibility and was weak enough to take on the job in the first place.

This is a lump of fool's gold as the organization thinks it is doing something sensible but really it is screwing itself up. The organization will probably blame the project manager for achieving so little and may even consign said manager to the dole queue which is very unfair.

In a balanced matrix the two groups (project and functional managers) are set equally in terms of authority. This is very close to the full-time assignment matrix as the resource manager looks after the resources' professional career and training and lends the resource out to a project manager who gives instructions regarding day-to-day work.

In the heavyweight matrix the project manager has a great deal of control over the resources, perhaps excepting only the decision about their future employment with the company. Whilst a resource is on loan from a resource manager to a project manager, it is the latter who decides on all matters technical, organizational and procedural.

The extreme case of the heavyweight matrix is where the resources are assigned to join a separate project team for the duration of the project. Here the resource manager acts much like a taxi firm and has similar authority over the people sent out to do work. The taxi firm (functional department) get the driver (resource) to your door equipped and ready on time. You (project manager) tell the driver where to go, how to get there and set the target timescales. It is a good thing that taxi firms don't run projects.

## Case study: MIRA

There are other ways of achieving a balance and the following case study which originally appeared in *Project Manager Today* and describes The Motor Industry Research Association's approach to the matrix issues.

## Admire a MIRA

It is probably a great deal more fun to be driven round the banked track at MIRA than to be in one of those poor cars that get smashed into concrete blocks. I am not quite sure as I have only tried out the former and that was about as stimulating as anything you can do with your underwear in place.

At first, The Motor Industry Research Association does not seem too likely to be a hot bed of project management but it sounded like a really interesting place to visit. Project management turned out to be a major issue at MIRA which at least means I can claim my expenses.

The motor engineers at MIRA spend their days subjecting otherwise peacefully resting vehicles to extreme conditions of all kinds. They whack them about the test tracks which include wet cobblestones, Belgian pave, Australian creek crossings, spoon drains and even New York sunken manhole covers. They smack cars into concrete blocks and steel barriers, heat them up, put them in simulated duststorms and bombard them with radio frequencies to try and upset the electronics.

If you happen to be a car, bus, truck or van and you find yourself being driven down the A5 near Nuneaton, break down immediately as you might be heading for a very bad day.

MIRA is run like a golf club. It owns its own land and has members who pay subscriptions. MIRA pays no dividends to its members, preferring to reinvest its profits in research and capital projects. There are greens but no clubs and no 19th hole.

Vehicles come from all over the world in complete security as Keith Read, who looks after client relations and is himself an ex international rally competitor explains. 'Manufacturers need to keep their cars secret. Perhaps it is months before a new car launch, perhaps for political reasons. We might easily have the next Escort and a new Rover next to each other in screened-off preparation areas, waiting for impact tests.'

Certainly there were any number of cars with varying levels of disguise circulating the facilities, wrapping your prototype in black plastic bin liners seeming to be the favourite camouflage technique. I searched the bushes for motoring magazine photographers hoping for an exclusive on a new car. I found a hare.

MIRA is open 365 days per year and deals with customers from around the world. 'What does Christmas Day mean to a Muslim?', asks Keith. It is a centre of excellence recognized around the globe. MIRA even lend their expertise to countries setting up their own testing facilities. MIRA consultants have been seen working in Hungary, Korea, China, Taiwan and throughout Europe. Competition comes from USA and Japan.

Before being allowed out onto the test track you have to check in with the equivalent of air traffic control and very strict rules try to keep the speeding cars away from each other and any unplanned encounters with hard obstacles. Until, that is, the time for crash test arrives. They test more than vehicles at MIRA with lampposts, kerbs and other bits of street furniture coming under the hammer. They had just finished testing a new motorway barrier design and if you are interested, this is how the barrier gets its comeuppance. If you have a nervous disposition miss the next paragraph.

MIRA start by building the barrier in just the way it would be built on the M25. Then it's down at the local scrap yard where they purchase an MOT-failed but intact large saloon like a Rover SD1. This is hooked up to a wire rope which passes around a pulley set in the ground near the new barrier and back to a huge winch machine. The winch is wound up slowly and released very quickly.

The high-speed cameras are set rolling, the warning sirens sound and the winch is released. The car accelerates faster than a drag racer and is chucked at the required angle into the barrier at the speed limit 100 metres from where it starting moving. On arrival it proves that there is no such thing as an immovable object. The path taken by the car post impact (excuse the pun) is filmed and the damage to the car is inspected. Then it's back the lab to develop the 1000 frames per second film and to write up the report. The scrapyard owner gets his Rover back.

One area of testing has less impact but was a surprise to me. Most modern cars use complex electronics to control everything from speed to brakes and these systems must be immune to all common forms of Electro Magnetic Frequencies (EMF). It would not be useful if every time you drove past someone on a car phone your car accelerated or cut out. At MIRA they have two opposing test facilities. One building completely screens its contents from EMF and the other provides no barriers to such emissions. You can drive a car into the first with its radio on and hear the radio go quiet as soon as the special doors slide closed. Then you can bombard the vehicle with all sorts of frequencies in a controlled way to see if anything odd happens. In the other one they had an excavator and were measuring the emissions 30 metres away. A controlled environment is at the heart of much that MIRA do.

The project which I had gone to see was the environmental test facility known dramatically as the Climatic Wind Tunnel (CWT). In this wind tunnel you can simulate a car driving from Siberia to the Sahara against a gale. You can pretend that you are whizzing at very illegal speeds down the M25 in mid summer (55 degrees, 95% humidity and 200 kph) and stop at a traffic queue to see if the engine boils or the air-conditioning fails.

The project manager for this new facility was Robert Birkmyre and he was appointed very early on in this process. It is a mystery to me where projects (and babies) come from and I often ask project managers for their views. At MIRA they get an idea for a new test facility and try to sell it to their own 'club members'. If the manufacturers are happy to buy a reasonable percentage of the facilities usage then the project will go ahead.

Robert sees himself as a slight maverick within the organization as his work involves many departments doing work that is outside their normal daily role. Robert devotes time very early on to 'getting input from the manager of each area, discussing their input, identifying specialists and getting a team together'.

Eventually there will be an order, either externally generated by a client or internally by the MIRA management, for the project to proceed. The project manager has authority from that point on – 'Anything that does not go through the project management does not happen', explains Birkmyre.

The project management for the CWT drew together a team of civil engineers, aerodynamicists, instrument specialists and IT people. The IT people look after the huge computer system that controls what will happen inside the chamber and that measures what happens to the poor old car.

Robert Birkmyre is a full-time project manager running a number of projects. Project managers are selected usually because of their backgrounds. There are civil engineering project managers and mechanical engineering project managers like Robert.

The project manager asks for specialists from the functional departments. 'I have requested an aerodynamicist from the Aerodynamics Department and I have asked for the best', explains Robert, 'and I have been allocated Geoff Carr for a period. That is now in my schedule. We keep talking to people and keeping each other up to date. If I asked for an engineer next week they would go up the wall; because my request was in good time I got who I wanted and everyone is content.'

Now there is an interesting resource-prioritisation system – the earlier you ask, the better choice you get.

'Of course things go wrong but we rely heavily on the high degree of motivation here at MIRA,' adds Keith MacKellar, General Manager of the civils, facilities and site, 'we pay people to do jobs and not just to attend. We have a great team here and often have to force people to take their holidays.'

At MIRA they use critical path analysis based on the Primavera planning tool. 'Planning suits us', says Robert, 'as it is a logical process. We produce network diagrams for all significant projects and then produce barcharts on a plotter selecting tasks for each department. We often group the work by contractor and then summarize the whole thing into a small chart.

'We do not use a planner as we see Primavera as a project-management tool and therefore the project manager is the user. We have fortnightly reviews and update the plans the day before the review meeting. At the review meetings we hand out the new plan, highlight and discuss the problems.'

Robert was sounding enthusiastic about his role as a planner so I asked how he felt about the planning function.

'I enjoy it', he surprised me, 'the computer system drives people to go on improving the plan but you have to stop somewhere.' He thought of three reasons in answer to my question about why he enjoyed it: 'It is a mental challenge, it is a change for me and it is still quite new to me.'

Robert uses three calendars on his plans – seven, six and five days per weeks – as some jobs like concrete drying go on every day and then contractors and MIRA people have varying working weeks.

'One day I would like to see all of MIRA's work planned in this way.' Robert thinks ahead, 'we are in a learning exercise, gaining expertise all the time. I often think how do I do this, how do I do that. We did go on the training courses that were part of the software purchase and we found the user support hot line very useful.'

The biggest surprise to me, which was not a surprise to MIRA at all, was their overall attitude to work and working on projects.

'We work for MIRA', explains Keith MacKellar. 'If MIRA does well, we all do well. Each department is a cost centre with a budget and target. Each department owns certain resources like test equipment and specialists and rents these resources out to outsiders and other departments. We don't have witch hunts here – when there is a problem we say that you have made a mess, what are we going to do about it?'

This all sounds too good to be true but my day with the people at MIRA seemed to confirm that this is the case. How is it that one organisation can achieve this level of co-operation and motivation whilst so many others are full of empire builders and people who are so afraid of doing something wrong that they do nothing at all?

And what is it that people in large civil service organisations fear so intently? You must have met the sort of people I mean – they are in a dead safe job where the ways of getting sacked are few and far between. Yet still they will do great damage to the company or to a project to cover up anything that might possibly be thought of as a mistake. In some companies people openly admit to errors so that everyone can do their best to deal with them and so that second occurrences can be avoided or at least minimised. But in the civil service type organisation it seems vital to maintain some appearance of not being at fault however much reality has to be stretched.

The civil servant type's first reaction to any problem is 'It's not my fault' and the reaction is never 'What can we do about it?'. This is particularly appropriate to project management as, in the words of Martin Barnes, 'You can only manage the work that remains to be done.'

So many organisations spend so much time worrying about fault allocation. At MIRA they do seem to manage the work that remains to be done.

They do not seem to have these problems at MIRA – there seems to be an atmosphere of openness which goes hand in hand with a general feeling of well being. They seem to enjoy their work. I would be green with envy if we didn't enjoy working for *PMT* at least as much.

## Project teams

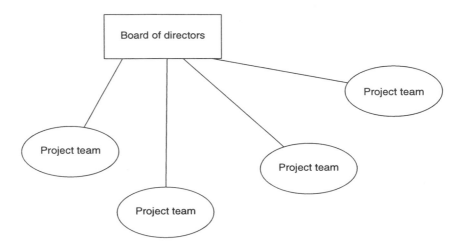

In the project-team approach teams are set up to manage the projects on a full-time basis. Normally these teams are given wide-ranging authority and it is essential that they are given clear objectives. The strategy is to give the teams clear objectives, substantial authority and leave them to get on with it. Highly-motivated teams tend to flourish in this environment, however, some senior managers have problems allocating sufficient responsibility to the teams. Also prioritization for resources between the projects tends to be on a 'survival-of-the-fittest' basis as there is often no overall guidance to resolve resource-priority conflicts.

This team approach is behind many of these 'flatter' organizations you hear about. The management replace the deep hierarchy of the organization as far as possible with project teams. Control is the aspect that suffers and the team approach is not ideal for those organizations where justice must be seen to be done. Such teams can be highly effective.

Organizationally, when the teams approach the end of their projects, new challenges must be found. Sometimes the team doesn't really run a project at all, but manages a department or a function.

## Project-support offices

Many organizations establish a project office to centralize some of the project-planning work. Within the project office are planning experts who spend a great deal of time planning. Any project manager can request plans, budgets and progress monitors to be carried out by the project-office members on a specific project.

The major benefit of this approach is that a considerable degree of expertise will be available to help plan each project and a high degree of standardization can be expected. The project office often produces reports for senior management on all the projects in their hands. The project office becomes a service department to the project managers, loaning planners out to the project teams as and when required. A disadvantage is that the project team do less 'thinking ahead' about their own project, effectively subcontracting that role to a professional planner.

Some programme-management organizations have a central planning office and call it a war room, project room or a visibility room. In these rooms all planning work is done by a dedicated teams of planners supporting the various projects and functions.

A war room has a lot of plans on the wall. They might be produced out of a project-management software package, they might be one of those flexible but mechanical wallchart display systems. There really ought to be an electronic version of a wall chart.

There are other uses of the term 'project-support office' and we had better make sure that you understand some of them. This office can be a more educational service in that it tells other people how to manage their projects rather than providing planning services to the various project teams.

Especially in IT, the project-support office is seen as a group who support the project teams as a sort of internal management consultancy. They offer advice on topics such as:

- project definition and justification;
- risk management;
- methods for monitoring and controlling the project;
- passing on advice from previous projects;
- cost-control mechanisms;
- reporting procedures.

In this sense the project-support office does not perform this work for the project teams, it tells the teams how they should do their work.

## Project boards

Where a number of projects are in hand within an organization there is nearly always conflict amongst the project teams for precious resources. Teams fight for the resources they need to complete their project but do not have the benefit of the broader view of the work. The broader view permits understanding of priorities between the various projects.

To help in this respect some companies establish project boards whose role is to act as a referee between the various project teams. Project boards expect a standardized form of reporting from all project managers and all project managers report to the board. When questions of priority arise the project board is aware of all the workloads and the conflicts and is able to make decisions about resource allocation.

Project boards may also set and maintain standards with respect to project-management methods, bringing a degree of standardization to the project culture.

## A master plan and who controls it

I have this vision of a programme-management environment and it looks like this:

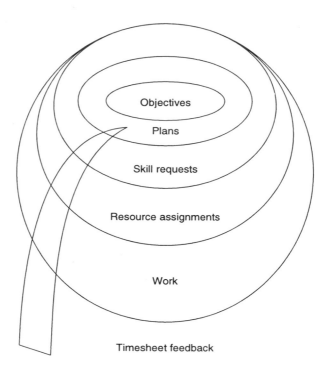

This starts off with objectives being the central and driving force behind a programme. Planners plan tasks next to the centre, sending out requests for skills. The plans show demand for printers, designers and engineers.

The resource managers substitute actual people or non-human resources for these skills. So the next ring shows the resource managers meeting the skill requests by booking people to tasks and assignments. Ideally this acknowledges the individual's expertise at a specific skill. You may be 120% efficient as a draughtsperson but only 85% effective as a CAD workstation operator.

The next ring, moving ever outwards, shows people doing the work and then the outermost ring shows timesheets being prepared. This outer ring loops up and over back into the central planning ring.

Within a framework like this authority is a big issue. If there is one plan for the workload of the company, who is authorized to change what? Who plans the work, who assigns resources, who collates plans? Biggest of all, who decides which project gets priority over another and who tells the system of these decisions?

An ideal programme-planning system would allow you to model the authority structure of the organization so that you could encourage or enforce the authority system of your choice. You could give each person on the system authority over certain elements of the plan. Here are some examples to make this point a little clearer:

- A project manager can work with her own projects but no one else's and she cannot assign resources.
- A functional manager can assign resources to tasks but cannot alter the timing of those tasks.
- The programme manager can allocate projects to project managers and only he can set baselines for later comparisons.
- Project managers can request skills but cannot change resource availabilities or assignments.

## Implementation problems

It is not easy switching to a programme-management structure. You buy system and install the software and start the system up. Once you have figured out how to drive the software you can start to add projects onto the system. At first most projects are not on the system therefore it only deals with a small part of the workload.

You really don't see the benefit, probably because there isn't one at this stage. As there is only a part of the workload on the system, the functional managers have to deal with requests of three kinds:

- the normal non-project workload;
- the projects on the system;
- the projects not yet on the system.

You have to find ways to encourage more and more people to get on board. The simple way is to get all new projects onto the system. This means that as old projects wither away and die they fall out of the picture and as new projects start they are added to the new computer system. Perhaps a critical mass is reached after a certain percentage of work is on the system and everyone suddenly wants to be on.

Certainly your case will be helped by a directive from above. Directives like 'Resource allocation priorities will go first to those projects on the system' or 'Bonus payments will only be calculated by reference to timesheets submitted' are hard but marvellously effective.

Either way you need a champion in the organization's senior management. This point has been made and stressed by almost everyone I have spoken to including BT, Inmarsat and the NRA. This high-level champion will help to push the idea through. You can become very dependent on the champion, so what if said champion leaves?

Some organizations run pilot schemes to test the ideas and these can be run successfully in certain circumstances. Pilot schemes are impossible or impractical if you decide to purchase a major-league heavyweight programme-management system. The cost of the system is going to be huge and there will be a team of people to look after the system and deal with user support and problems. I call the purchase of a heavyweight system the big bang approach as it is not unlike the formation of the universe.

> And the one who formed the universe said 'Let there be stuff.'

The alternatives involve some kind of lighter-weight system being installed piecemeal through the organization over a period of time. This is sneaking in through the back door. There are 'consolidation' tools that have the stated purpose of collecting and assembling project plans from many projects and this is also a way of avoiding the big bang.

Behind an implementation you might expect to find a working group or a steering committee with representatives from many areas of the company, including the board. This group represents the various interests around the company and is charged with getting programme management up and running within the organization. Each member represents some interest group and reports back to those people between meetings. Such groups should be empowered to select the tools and look after the staff requirements of the system.

## Case study: BT Labs

The following case study examines matrix management and the multi-project management approach adopted by a division of British Telecom. It originally appeared in *Project Manager Today* and shows how one group of people met a range of problems in setting up a programme-management environment.

I am grateful to BT and especially Dot Hurley at the BT Laboratories near Ipswich as well as *Project Manager Today* for permission to use this case study.

# *Managing the matrix*

Once upon a time project managers worked on one project at a time.

The history of project management sweeps majestically over great bridges and dams and through major tunnels. It is powered by huge power stations and grand hydroelectric schemes.

Today things are very different. Most projects are run within an organization running many simultaneous projects in what is called the Multi-Project Environment. People involved in running many simultaneous projects talk about a project-management culture. Programme Management seems be emerging as the common term.

Not surprisingly the need for techniques and tools in programme management is very different to that of the single-project environment. I hope that in the next century we will look back on the nineties as the period when project management came of age as tools and techniques appropriate to the needs of running many projects were developed and became a part of everyone's normal management tool kit.

Many companies have tried hard to create their own systems for dealing with many simultaneous projects. I have observed National Rivers, IDM, Research Machines and many others whilst they think hard about the way many projects can be managed and controlled. The conjuring trick involves balancing the needs of the many projects with the needs of the functional departments and their resources. Such companies often take ordinary software tools and modify or extend them to solve their needs. This month's tale is of one such organization who have kindly made their experiences available to all through the pages of *PMT*.

### The environment

The generous organization we are today concerned with is the Programmes Unit within the Network and Services Development Directorate of the operating division of BT known as Development and Procurement. If I don't reduce that mouthful to D&P you will be reading all day. Or not.

Nation-wide BT have a trunk network (the telephone equivalent of motorways) plus local networks (roads and streets) connecting your local exchange to your home, office and Aunt Mildred. There are many different types of links especially in the international trunk network: satellites systems, microwave links plus undersea and land cables.

This nation-wide mass of cabling connects exchanges within BT and larger corporations together and it is inside these switch rooms that big changes are going on. The idea of a switch room full of relays clicking away is a thing of the past; today's switch gear is made up of lots of boring grey computer boxes. Those enigmatic clicking relays have been replaced by little lights that flicker on and off – less exciting but much more reliable – all because the 1990's exchanges, PABXs and International Networks are all software driven. Your local exchange is probably a standard BT box controlled by a VAX or UNIX minicomputer running some specialist software written at the Martlesham Heath Laboratories of D&P near Ipswich.

The people at Martlesham are in the business of developing new IT products and services for sale. Usually one of the BT operating divisions will come along with a need, asking the labs to design and build a solution.

This part of D&P is therefore essentially a software and hardware house. A strong client/contractor relationship exists between the various divisions and departments of D&P and everything is done as a project.

They have competitors. The client division of BT could go to another software house or Telecom supplier. D&P must prove that they can perform.

*At the labs*

The Network and Service Development programmes use a matrix-management approach to link both functional teams and projects.

The main functions that you would see across the top of the matrix are shown in the table opposite.

Down the side of the matrix is an organizational structure of projects people.

Most of the work is project managed by the Network & Services Programmes Unit. Dot Hurley works in this unit and was our guide to her world of programme management in which 75 people form the 'project' side of the matrix. There are Programme Managers looking after major programmes, each of which contains a number of projects. These projects are likely to be linked by type of work or client.

**Table 4.1** BT D&P functional departments

| Department title | Function |
| --- | --- |
| Business development | tout for new business, select the workload |
| Requirement capture and system design | define requirements and design systems |
| Engineering areas | build the products – software |
| Validation, Verification and testing | testing |
| Supply and Support | installation and in-field support |
| Programmes Unit | project management and control |

Each Programme is managed by an Executive Board. Whilst a board member might hold more than one role and the board would look after more than one project you would expect to find these roles represented within the board:

Programme Manager
System Engineering Manager – the design authority
Engineering Manager or Implementation Manager
Validation, Verification & Testing Manager
Quality Manager
Release Manager
Supply Manager
Operations Manager (project control)
Major Resource Manager.

Each Programme Manager would look after a number of Version Managers who oversee development projects leading to new versions of the software systems for a particular need.

*Across the matrix*

These Version Managers effectively subcontract assignments to the resource centres. The functional department have the resources that do the work so they contract to do work for the programme boards. This is sometimes called an Assignment Matrix.

Some of the advantages of this approach are that the specialists have a career structure within their function and the 'two boss' problems that can arise in some organizations are avoided. On the other hand there are no complete, devoted and motivated project teams working towards specific goals.

There is also an operations resource pool made up of people who do project control like planners and administrators and these support the programme boards.

The Programme Managers meet monthly to discuss inter-programme issues like resource conflicts. They worry about getting 300 man-hours of work out of three engineers over Easter long before the chocolate eggs go on sale, not on Good Friday when it is too late. Issues that don't get resolved at these meetings go up to departmental management meetings which bring together both project and resource management. Together, the two sides of the matrix have the power to make decisions.

Before the drive towards programme management there were hundreds of projects in the resource-based functions. This lead to lots of duplication as similar products were often developed simultaneously. Whilst some departments executed whole projects, some sections specialized and offered their services to other departments.

Customers in the past complained of different reporting techniques from each area of the business and the lack of a single point of contact. Now every customer has a person to deal with and reporting is being standardized.

### Changing the culture

I met Dot Hurley at BT's huge Martlesham Heath laboratories outside Ipswich. It is the sort of place where they check you in at the gate and give you a map to help you find the right building. It is like space: big.

To meet Dot you find the right building then find the right corridor and finally find the right room. She is worth the search and knows her stuff. She affirms, 'Everything here is a project. The only way to get funding for anything is via the project mechanism.' It's a good thing that the tea is free.

Dot was given the task of improving planning some three years ago. She was a software-development engineer building an Accounting and Management Information System. 'I used to plan my projects', she remembers.

'It all started off as a divisional slush job – we decided not to look at just planning but to look at project control. There was a team of two in the early days looking at feasibility: what needs to be looked at, what do we cover?'

So Dot's team of two set about designing a planning and control system for day-to-day project management.

Other demands came along because of the introduction of quality management and the desire for the network management department to achieve BS 5750 qualification. This meant drawing up procedures showing how projects should be managed and controlled for the quality system.

**Project-management systems**

When Dot got involved in the idea of a programme-management system Hoskyn's Project Managers Workbench was being used extensively. Dot felt she needed a tool that could bring together many projects and provide an overall picture – a programme-management tool.

She recalls, 'In the early '90s not much was available. There were half a dozen big systems and loads of PC tools. Eventually Cascade was chosen to connect and link the many PMW plans.

'We have PMW and many people are happy with it. The problem was to complement PMW with an overall view. Cascade sits above the individual plans all of which are held on our many copies of PMW.'

Today Cascade runs across a number of UNIX fileservers and everyone involved in project control work works on a networked workstation or PC which is hooked up to the Sun UNIX fileservers.

To get a software system working on many projects often involves some form of computer network and this is the growth area in project-management tools. BT, as you might expect, take inter-computer communications for granted so the assiduous reader must assume good communications across a number of computers perhaps running different operating systems.

The Programme Managers and Version Managers create the work breakdown structure – an overall plan for each work element or project. They decide which functional department should work on each element and negotiate with the appropriate resource managers. The functional managers plan their work in detail on PMW.

Yes, it is the functional managers who plan on PMW the work load passing through their hands, not the project teams. Plans cover the function's workload in detail, resource by resource, and deal with their complete workload across all projects within the department. Each plan is cut up and imported into a Cascade 'Cost Account'. A Cost Account is Cascade speak for the part of an element of work being carried out by an organizational unit on a particular work package. Look for a place where the OBS and WBS intersect and you will find a Cost Account.

Therefore the PMW plans relate to the top of the matrix – the functions – and the Cascade plans form the side of the matrix – the projects. Connections from one project in one organizational unit to another are handled within Cascade. The Cascade workload provides the agenda for the monthly meetings.

PMW allows plans to be created at Phase, Activity and Task level. The whole detail of a PMW plan is normally summarized into

Cascade at the activity level although this depends on the project and the level of detail required.

On the first working day of every month the updated detailed PMW plans are sucked into Cascade. This could lead to chaos manor if the plans being drawn in do not have some consistency amongst them. The Project Management System Group lays down a planning method which gives the planners freedom but does set some standards, for example all work is planned in man-days. A version manager might impose task-naming conventions and there are standard templates available.

## Implementation

At the moment [June 1993] Dot's team – the Project Management Systems Group – have PMW and Cascade happily working together. They are currently very much in the implementation stage.

The first step is to get PMW accepted more widely. Some long-term users saw the benefits of Dot's approach but some of the more old fashioned types were less open. Dot had to cajole and bully until last autumn the Resource and Departmental managers realized that they didn't know who was doing what. There was no clear picture of resource availability. Some bright spark suggested they needed a resource plan and someone suggested a spreadsheet. Dot got involved and she suggested planning and persuaded them to plan as a means to get what they wanted. She was nearly ready with the tools when the managers realized they had a problem. What timing!

Not everyone is mandated to plan. Everyone is encouraged to plan on PMW, every section (30–50 people) will have at least one plan showing all the work.

Dot recalls, 'It has taken from October to February to get that going throughout 40 groups. Some didn't have PMW, some did. We went to see everyone and asked to see their plans. It helped enormously to have direct management backing. We tried holding hands and per-suasion but we do now have a last resort: the black list.'

Dot uses the Supply and Support department as an example. 'They have been able to prove their overloads since they started planning and tracking. Their plans are relatively simple with only two or three tasks per "mini-project" but they see the benefits. The whole section turned about from being very "anti" and became very positive.'

### Training

Clearly with many people adopting a project-management culture and using either PMW or Cascade training is bound to be an issue.

Dot explains; 'We organize PMW courses on site modified to suit BT's needs and run by Hoskyns. We currently have about 200 users.

'The majority of Cascade training is done in house by Mantix. These courses are aimed at the operations managers who are the key users. We have trained about forty people to date.'

Dot has become a crusader. She spends time giving talks and presentations on what the system is all about to the senior people. These talks are not detailed but at a concept level. Her team tailor-make training course for the resource and programme managers due to the varying level of use. She reminds us: 'some managers get their hands on the systems and many engineers are very computer literate. And don't forget that the network system is essential, electronic mail and departmental networks spread around the country. WAN and LAN familiarity helps a lot.'

An obvious question when talking to someone who is in the midst of implementing a project-management culture in an organization this big is 'If you had to do it again, would you do it again differently?'

'I would have resigned' is her immediate reaction, then she reflects. 'Stronger and more visible backing from senior management early on would have been a huge help. You must have backing even to sell a good idea. We began with a softly softly approach, trying to persuade people but it didn't really work – we had to tell.

'Also I wish we had started with a clean slate and fixed goal posts.'

Dot is intensely proud of her team who, judging from the cartoons and other paraphernalia around the offices, have as good a time as you can have at work. You get the idea that the team has made the changes possible and enjoyed themselves whilst doing it.


### The future

Dot's team is bringing earned value analysis into the picture in an attempt to replace the current wadge of paper reporting progress with two simple graphs. She explains, 'we are using a Red-Amber-Green (RAG) status system but there are no set rules. I would like to replace this arbitrary, subjective system with a scientific mathematical calculation giving RAG status on both time and cost for every project.

'Right now we produce a standard report for every programme on one page plus RAG status for each project manually – but this will be automatic one day.

'The whole system should be up and running by 1st April and will need some time to let it bed down. We do have an improvement project plan for this project.

'There is a timesheet system in hand as well. This will sit below the PMW level and it will be necessary to impose a consistent system for this. One day everyone will plan with PMW and the plans will be updated with actual work done by assignment.

'You could take this information into Cascade and from there to other central systems like accounts and therefore directly raise invoices. Under such a regime if work was not planned it would not be bookable and there would be no income from it.'

Judging by the level of interest at the recent seminar on Adopting a Project Management Culture at PROJECT MANAGE-MENT '93, programme management is a key topic amongst project managers throughout the land. At that event three speakers from TSB, British Rail and Computervision all talked about their experiences as they moved towards a programme-management culture. Ms Hurley has kindly shown us the BT way of doing things, some of her scars and some pitfalls to avoid. Thank you, Dot.

We hope her system spreads throughout BT and that the team stays in gainful employment. After all, we contribute to their salaries every time we pick up the phone and they deserve good jobs.

## *Roles within the matrix organization*

### *Introduction*

What I'm going to try to do in this section is to outline the roles that exist within the generalized programme-management environment. When I say 'roles' I don't mean ham rolls, Rolls Royce or rolls in the hay. I do not even mean individuals, rather the actions and responsibilities they adopt.

Many organizations cannot afford a full-time project planner. So someone, usually the one person who actually has figured out how to make *SuperWorkPlanner* version 9.6 produce a barchart, has the planning role for a day or so each week. She might be a design engineer but on those day each week, she has the role of planner.

In larger organizations there may be many planners and they may all be full-time staff. But don't forget one person may carry out many roles, one role may be carried out by many people. In some organizations and textbooks these roles are known as logical functions or shared functions.

 I'm doing this so that you can think about your own organization and see where these roles appear or should appear. It makes some sense to read the section about organizational structures as well as this to get an idea about those as well. You might use some of this as script for your own internal reports so I'll be a little more serious – it's going to be hard but I'll try.

Before getting into specific roles let's take a look at the overall picture. This diagram shows a fairly typical matrix-style structure – the sort of thing you might find in any organization.

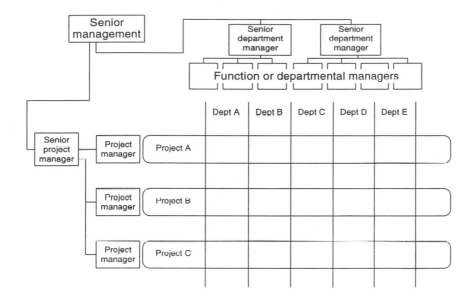

Within this structure there are four major roles to be adopted by someone or some people. They are:

- project managers;
- functional managers or resource managers;
- resources or operatives;
- programme managers who act as referees or umpires.

Within the structure of a matrix the project managers push, pull and coerce projects across the matrix. They may have no resources of their own but they do have projects to get done. They get these projects done by getting the functional or departmental managers to allocate or loan resources to their projects. Across the top of the matrix are the functional managers and they have a team of people who provide a specialist service.

The senior management function sits above and around both axes of the matrix. The senior management hire, direct and fire the functional managers and the project managers. These people understand the priorities of the organization and have authority over the project and departmental managers. Therefore the senior management will act as umpire or referee to settle priority issues when the demands of many projects overload a functional department's ability to provide resources for a period of time.

ces or operatives work within a functional department and the projects that pass through their hands. Projects are ss the matrix, spending time in each department as they get )wards completion. These resources are allocated to work ts in a variety of ways. Sometimes they are seconded to the full-time or part-time basis, sometimes the department ᴄᴏɴᴛʀᴀᴄᴛꜱ ᴛᴏ do the work.

Most organizations do not have a 100% project workload. Any programme-management organization is likely to have a continuing background workload of ordinary non-project work. The resources will be involved in project work and a whole raft of other work that does not conveniently break down into projects.

The background workload might include maintenance of equipment, looking after users of old projects, selling activities on new projects and changing toilet paper in the washroom. They all involve doing work and take time but are not connected with the project workload.

Now let's take each of these four major roles one at a time in a bit more detail. I'll take these four roles and for each I'll describe the sort of responsibilities they might have.

You might notice that some of the areas of responsibility are optional. For example, some functional managers exist to loan resources out to project teams whilst others exist to perform work for a project manager. You should be able to select the responsibilities that seem appropriate to your organization.

## Programme management

### Brief description of role

- The management of the enterprise's workload being made up of a number of projects and continuing work;
- the delivery of a cohesive set of external projects which collectively meet the organization's customers' needs;
- the delivery of a set of internal projects which are collectively aimed at meeting the organization's objectives and business strategies.

### Responsibilities

- Defining programme goals based on the organization's strategy;
- continually evaluating potential enterprise-wide benefits of internal projects to justify the programme in terms of the changing corporate objectives;
- evaluating risks and benefits from external projects;
- describing the scope and boundaries of the programme;

- maintaining relevant section of organizational breakdown structure and work breakdown structures to adroitly fragment the workload amongst the project-management team;
- controlling the content and membership of the programme- and project-management team;
- establishing and operating an approval procedure by which projects can be checked and approved;
- initiating the programme infrastructure by gaining approval at board level to embark on a feasibility study for the programme;
- approving projects' budgets and timescales in appropriate detail;
- assigning projects to project managers;
- managing conflicts between projects and between functions;
- monitoring progress against project milestones or in appropriate detail;
- experimenting, modelling and making decisions about future work-load and opportunities;
- working with potential projects and workload;
- considering strategic implications of each project;
- understanding the impact of one project on another;
- approving changes in project status, e.g. giving permission to start a project or a phase;
- maintaining library of standard projects and outputs;
- ensuring projects are formally closed;
- analysing past project performance and monitoring feedback systems;
- risk management.

## Project management

*Brief description of role*

- The management of specific, named projects.

*Responsibilities*

- Driving a project through the enterprise;
- planning what tasks and products are required for a project;
- estimation of work content and skill requirements for tasks and products;
- negotiation of timescales with resource managers;
- deduction of the project's structure, e.g. work or product breakdown structures and PERT models;
- receiving and utilizing resources 'seconded' by the resources managers;
- subcontracting packages of work to the resource management;

- monitoring progress and adjusting timescales of key points in the project if necessary;
- defining deliverables of tasks and projects;
- estimating budgets and timescales for projects;
- gaining approval for budgets and plans.

## Resource management

*Brief description of role*

- The management of a number of specific, named resources

*Responsibilities*

- Recording expected availability of individual resources (holidays, overtime, etc.);
- allocation of skill levels (capabilities) to resources;
- allocation of specific resources within a department and/or enterprise to satisfy the requirements made for certain skills by project managers and others (the resource manager 'satisfies' the need for, say, 100 designer hours by allocating specific designers to the task);
- secondment of resources to a project manager for a period of time (in a secondment matrix);
- performance of tasks as requested by a project manager (in a sub-contract matrix);
- prediction, communication and resolution of resource shortfalls;
- responsibility for maintenance of specific local area of the organizational breakdown structure;
- negotiation of timescales with project managers;
- optimization of resource utilization;
- maintenance of resource output tables and database of production outputs;
- instruction of distant resources on their next work;
- ensuring timesheets are completed on time;
- working within departmental budgets.

## Operatives/resources

*Brief description of role*

- Anyone who actually does direct work towards the project's goals.

*Responsibilities*

- Performing work on the tasks within the projects;
- reviewing work done by others;
- measuring and reporting on actual achievement, work done;
- updating estimates of work remaining;
- committing update information to the system;
- comparing remaining work with work planned;
- comparing work done with work planned;
- receiving instructions on future work;
- enquiring on the history of work completed;
- understanding how own work fits into overall plan.

There, that wasn't too bad, was it? You have my permission to impress your fellow workers by using any part of the above in documents internal to your company.

# Extra baggage

## Methodologies: PRINCE and PRIDE

It was Maureen Lipman who, in a BT ad, brought scientists and engineers down to earth with a breathtaking bump when she described her grandson's 'A' level results: 'He got an ology, already!' If you feel the need to make something sound important, significant and above all, expensive, give it an ology.

I'll talk about methods, not methodologies. I suppose I could go through this chapter later and globally change method to methodology but I have found global change to be a dangerous tool and between us we don't need ologies.

A method is simply a systematic way of doing something. Methods lay down the stages, phases and procedures by which projects should be managed. If your organization likes to have things done by the book, the methodology is the book by which everything is done.

Now if you write down the way in which you think people should run their projects in some detail you would have a nice book. Your chances of selling large number of copies of your book at huge prices would be low. *How I Run My Projects* by Fred McSmigginbotham at £300 would languish at the lower end of the bestsellers list alongside *Lesser Known Trams of Grantham* and *Ten of my Favourite Bricks*.

But if Fred called it a methodology, got it published by the government and gave it away free many people would use it. There may not have been anyone called Fred on the team but that is roughly what the CCTA did. The CCTA is the government centre for all things computing and they beavered away for yonks and came up with two things. One was a methodology (oops – I got carried away again) which we'll take a look at in a minute and the other thing was a stunning acronym for their new publication. They called it PRINCE which stands for PRojects IN Controlled Environments. You may give thought to which of the two took longer to develop.

When you come to choose a method you can have PRINCE more or less for free or buy a method all of your own. You can employ a highly-paid consultancy to tune PRINCE to suit your environment or create a new method completely from scratch. You can set some of your own people onto the task of dreaming one up for your organization but just before you do let's ask a meaningful question:

What does a method look like?

PRINCE is a book, well actually it is often in a ring binder. It is quite like those magazines that you buy in 12 weekly parts and that build up to a complete record of fighter aircraft from the Gulf War or, less aggressively, Javanese knitting patterns. Do you buy those mags? If you do give me a call, I've always wanted to meet someone who did.

Before reading this examination of PRINCE you must understand that the following is a very, very, very brief summary of PRINCE to give you an idea of what it and many other methods contain. Small people cannot peer over a complete set of PRINCE documents when they are piled on a normal desk, so be prepared. If you ask for a set of the PRINCE documents do not expect Postman Pat to drop them through your letter box, expect Pickfords to bring them round on a low loader.

Another big note – PRINCE breaks a project down into stages, each of which starts with a specific set of deliverables, documents and approvals and ends with other documents and deliverables. I guess that one of the prime tenets of a formal method is the breaking down of a project into phases or stages.

A significant part of PRINCE relates to this idea of breaking all projects down into discreet stages each of which has a budget, timescale and objectives. Very frequently, passing from one stage to the next requires the submission of certain documents and sometimes a business case to request permission to proceed into the next stage. Frequently the handover from one stage to the next involves the handing over of the project from one person to another and you would be wise to accept a project in these conditions only when you are satisfied that everything is in shape and hunky dory. Taking over a stage in a method-driven company is much like buying a second-hand car – look before you leap.

Here goes with a brief PRINCE overview. PRINCE deals with three main elements:

- organization
- plans
- controls

I'll take these three elements one at a time.

## Organization

The section on organization within the PRINCE method explains how the organization is structured and who is responsible for what. When I say 'who' I do not mean Fred, Sally or Azif. I do mean the roles that have to be adopted. PRINCE, recognising that people come and people go, sticks to roles. There may not even be a one-to-one mapping of roles and people. In small organizations one person can adopt many roles. In a large organization roles can be carried out by a whole department.

PRINCE recommends the establishment of a project board. The board manages the project from a fairly high level. Three roles are adopted by the board:

- executive:         corporate guidance and assessment;
- senior user:       representing the users of the project's deliverables;
- senior technical   representing those responsible for the technical aspects of the project.

Next PRINCE suggests there should be two project-management roles. The project manager plans the project, reports on progress to the board and initiates activities. A stage manager may be appointed for each stage of the project (design stage, installation stage, etc.) and is responsible for delivering the products of that stage and meeting quality, time and cost targets.

| Stage manager sounds very theatrical doesn't it? |

There should be, in the world according to PRINCE, a product assurance team. Made up of a business assurance co-ordinator, a technical assurance co-ordinator and the user assurance co-ordinator, this group has two responsibilities. One is to ensure that the project board can believe what the project manager is saying about the project and the other is to offer help and the benefits of experience to the project manager.

PRINCE also recommends the formation of a project-support office where the experts in programme management, project management and project planning on all projects within the organization can sit and share horror stories about the project workload. The business assurance co-ordinator and technical assurance co-ordinator roles can become specialist roles performed on all projects in this centralized project-support office. Representing users is likely to be more project dependent.

## Plans

This section talks about the planning documents that PRINCE people prepare or expect to be prepared. PRINCE talks about plans of various kinds and you may experience the difficulty I found when they used the word 'plan'. When I think of a plan I get a mental picture of a barchart or Gantt chart or perhaps a PERTchart. As you will see in a moment, some of the PRINCE plans are not likely to look like a barchart at all.

Plans start with the **technical plans**. These are traditional timescaled plans showing activities that will go towards achieving the project goals.

**Resource plans** are used to identify the number of resources and the type of resources required to achieve the project. The project resource plan is a top level plan for the whole project and the stage resource plan is a more detailed look at the resource requirements for a stage. These will look much like histograms in graphic or tabular form. Resources in PRINCE can be human, 'equipment and other resources required' or cost.

The PRINCE guides talk about appropriate levels of planning. There can be, for example, the following plans within the overall project technical plan:

- a stage technical plan for each stage in the development process;
- a detailed technical plan examining in greater detail some complex or major activity;
- individual work plans detailing specific people's activities within a stage;
- an exception plan is produced when things have gone wrong and shows what action is to be taken to deal with the 'deviation'.

Other plans can be broken down in a similar way.

PRINCE is keen on product-orientated planning. A key feature of the PRINCE philosophy is to focus on products rather than tasks. The idea is that you concentrate on the deliverables of the project and of each stage within it. This focuses people's thinking towards the product of the project and the part of the product achieved by each phase or stage.

The danger is that once tasks and small jobs are identified, it is easy to concentrate on achieving those tasks and losing sight of the larger picture in terms of what you are actually trying to achieve.

To develop this concept further, PRINCE talks about the product breakdown structure (PBS) which is much like a work breakdown structure (WBS) with a heavy concentration on what is to be produced (pp. 55–62).

## Controls

The are two forms of control. One deals with the documents that the project creates. These documents need to be controlled so that the right people have the right version of the right document at the right time. The other side to control within PRINCE is about the methods that the organization uses to keep a beady eye on the projects and to propose control actions.

Controls in PRINCE-speak are either a meeting or an assessment and we can look at three types of meetings and two assessments.

The **project-initiation meeting** is designed to be the final approval to go ahead with the project. It is the moment of commitment. Of course, the meeting could not sensibly assemble all the data in one session, so someone has the job of assembling a project initiation document which contains loads of information about the objectives, timescales, budgets and aims of the project.

I am very keen on this stage as so many projects get going with only the vaguest ideas of what they are supposed to be aiming to achieve. More importantly, each person will have a different idea and it is very likely that your boss's idea of what you are supposed to be doing is likely to be different to your own. Therefore, no matter how good a job you make, your boss will feel a little miffed. It is unfair but true. A solid definition of what the project is about and a formal signing off will reduce the chances of this happening. It will not reduce them to zero.

As the project gets going it passes into its first phase which might be a design phase or a detailed requirements-definition stage. You might consider a **mid-stage assessment** to provide a control point for the project-management and project board. At such an assessment the team compares progress and development with the original aspirations and comes to decisions. If things are going very badly the project might be shelved or scrapped. If things are going well the team might give the go-ahead for the next stage even though the current stage had not ended.

More often the project is ambling along reasonably normally and the team talk about the unplanned situations that have arisen and make decisions about dealing with them. These mid-stage assessments are 'optional extras' in PRINCE.

During each stage the team are likely to meet regularly at a **check-point meeting**. These are regular opportunities to review who has been doing what on the project in the last few weeks. At each checkpoint meeting the stage manager, helped by the business assurance co-ordinator and technical assurance co-ordinator, produces a checkpoint report for the project manager.

At the end of each stage there is likely to be an **end-stage assessment**. Here the project manager presents to the project board a report defining the current status of the project. He often seeks the board's approval to proceed to the next stage. Now get that idea stuck in your head. At the end of each phase or stage of the project the project manager has to prove the case to proceed with the project. It is a great way to keep those (us) project-management wallas on their toes.

Finally there is a **project-closure meeting** which closes the project and hands the deliverables over to the users. The meeting is supposed to devote time to examining the history of the project and learning lessons from it that can be applied to later projects. The team are probably far too busy sinking pints of real ale or sipping gin and tonics to celebrate the end of another project to worry about looking back.

That is PRINCE in a nutshell. The actual set of documents is voluminous as it goes into detail about the agendas for meetings, descriptions of the various roles and the contents of each report. You can take PRINCE out of the box or adapt it to meet your own needs. Adapting it can mean simply changing the names and titles used to suit those within your own organization. It can mean a thorough rewrite. The definition of stages is very likely to need tuning to suit your lifestyle.

Adapting PRINCE or starting from any other source is likely to be very expensive.

## Why do you want a method?

A method lays down house rules for running a project. It describes the roles that people should adopt, their relationships with other roles, the stages and phases that the project should pass through and the documents that should be prepared.

It is going to be at its least useful in a single project. By the time you have written down your method a single project would be nearly over and there will be nothing left to apply it to. Methods clearly are going to work where you have a number of projects to undertake.

That is not to say that a method is of no value on a single project. I met a project manager on a large, single-construction project whose team was made up of people from many organizations who, at the start of the project, were strangers to each other. They got the working team to set out a method because it forced the team to sort out a way of working amongst themselves. The value was partly in the preparation of the book and partly in issuing the book to the other late-joining team members.

Not only is the idea of a set of rules for running projects only going to apply in an organization where there are many projects, it is going to be at its best where those projects are similar. Not surprisingly, the whole idea stems from the world of IT where all projects are similar.

> There are some sweeping generalizations in this book but that is too much to handle.

Now listen here Mr Footnote, don't give me a hard time – in all IT projects there is a specification process where the computer people try to understand what the user wants; there is a design process where the IT people try to come up with a system that will meet the need and there is a build phase where the programmers write the thing. Then there is a testing and installation phase where the users discover how closely the system meets their needs. From a project-management viewpoint all IT projects are similar.

It is these phases that PRINCE and its commercial competitors aim to tie down. So choosing to have a method is like choosing an attitude to life. If you want to control as much as you can so that you get fewer failures (but no great successes) then a method is for you. If you want to give people their heads so that you get some spectacular successes (and a few equally spectacular failures) then don't bother reading PRINCE. On this basis you will not be surprised to hear that PRINCE is widely used in government IT projects.

There is a halfway house where you make methods tools available to all and allow people to select from a methods toolbag to suit their environment. This is the approach Inmarsat use to help bring some method to their wider variety of projects.

Another sound reason for acquiring a method is self-preservation. A set method will almost certainly allow you to point the finger of blame at someone who bypassed the method, someone who didn't produce a product breakdown structure or who skipped a phase review. And if everybody did everything by the book you can at least say you tried your best by using the best methodology around.

Yet another sound reason is so that you can answer 'yes' to the question: 'Do you use PRINCE or another approved methodology?' This question tends to arise when negotiating to do IT work for a government department. Many will only employ contractors who use such a method. Now that is a good reason for at least acquiring and understanding a sensible method.

If your organization does not have a method perhaps you might suggest that you investigate the subject more fully and then run a project designing and installing a method appropriate to the company. If this unlocks doors with government, you might enhance your personal reputation a touch.

It is worth thinking about who is best suited to creating a method for you. You have a number of choices here. You can recruit on a full-time or part-time or consultancy basis some methods people to come in, learn about your company and write a method just for you. Your own personal ology. Of course you will have to pay them for their time whilst they learn about your organization and the way it works. You'll also have to pay your own staff for the time they spend talking to the consultants explaining to them how the company works.

Another approach is to take some people within your own organization who are familiar with its systems and methods and get them to learn about method theory and create one for you. Rather than take a methods expert and teach them about your company, you take a company expert and teach them about methods.

Perhaps the ideal is a mixed team of both. One of the more risky approaches is to select a fine member of the organization whose current prime objective is to survive quietly until the retirement party ends and the carrot turning begins. As this person ran some really great projects in their hey-day and since they know the company inside out they should be ideal for the job, but their motivation is low and they may find extreme difficulty in understanding what this method is actually for.

Some organizations take a bright young thing who has recently joined the organization. Said bright young thing devotes some time to researching with a fresh eye how the organization works and makes notes in a procedures manual or a method. There is a spin-off benefit here apart from the method itself and that is you get a bright young newcomer who is familiar with all aspects of the company's machinations.

Methods tend to get stuck in a groove. As times change, as organizational structures change and as the company metamorphoses to react to those changes, the methods have to be kept up to date. There is a danger

PRINCE or **PRIDE**

that the method becomes the holy grail, the organizational equivalent of the law. People often blindly obey the strictures of the method's scripture long after it has become irrelevant to the company's current way of working. It pays to set up a group now and then to update the method in light of changing circumstances. I have never seen a company with a suggestions box for improving the method but some system for keeping an eye on the method reduces the danger of stagnation.

## A final word

One final word about methods and methodologies. This world of methods is packed with acronyms and I think that one of the best is the other popular meaning of PRINCE: Projects Running Into Confused Excuses. I also hear of a much more common methodology than PRINCE. It is called PRIDE – PRojects In Deep Excrement.

## Case study: Inmarsat

I am happy that Inmarsat allowed me to use this short article which puts their view of methodologies as well as other elements of programme management in perspective from an engineer's point of view. It first appeared in *Project Manager Today*.

---

## Space is big

---

Something very special happens 36 000 km above your head.

At that precise height satellites can be persuaded to stay more or less still relative to mother earth. Anything lower or higher will seem to move over the surface of the earth and it is very much easier to point your satellite dish at a non-moving target.

At 36 000 km a satellite is geo-stationary, circulating with the same 24-hour cycle as the Earth below. For comparison, shuttle Endeavour climbs to around 300 km and Jumbos cruise at 30 km.

We have author Arthur C Clarke to thank for the idea of the geo-stationary orbit as it is he who specified the whole idea in the 1940s. If he had patented the idea he would have become very wealthy indeed – the idea was tested in the '60s and is now in common use. Fortunately he has made so many TV programmes, films (*2001*, *2010*) and written such great books in both Science and Science Fiction (the Foundation, Cradle and Rama series) that he rarely has trouble paying the milk bill. It is you and I that have benefited rather than the inventor.

One of his rather less tested ideas is a geo-stationary space station with an elevator back to earth. The connecting cable would be a one-molecule thin cable 36 000 km long with a 'station' at each end. In *Paradise Lost* he wrote about this scheme figuring that the elevator would use very little energy as it goes up and down conveying bits of space vehicles economically into orbit. The problems of making a one-molecule thick cable seem no more daunting today than those of placing a vehicle at 36 000 km must have seemed at the beginning of the Second World War. Mr Clarke is one bright guy.

So is Keith Rowe. He is not as well known and probably not as wealthy but he contributes to our ability to speak to people and see events all over the world courtesy of that great global communication system known simply as 'by satellite'.

After graduating from the University of Canterbury Keith joined the RAF and was soon a Satellite Operations Officer looking after military space vehicles. 'For me,' explains Rowe, 'space was an accident. I looked at various RAF sites and asked for one "near space". I was always fascinated by sci-fi and space. I took over command of a telemetry station at Farnborough before going to Kinloss to look after Nimrod avionics. I didn't enjoy that one. Then I spent some time in a software-testing team. Therefore satellites seemed sensible for me.'

After time at another RAF telemetry station, in control of a defence communications network in Bath, Keith went into mobile satellite communications on the Falklands connections. 'I was fortunate,' he remarks, 'I didn't have to go to the Falklands'. After eight years' service he left the RAF early and joined Inmarsat.

At first he managed contracts, spending time on secondment to a French national space contract. Later he became Operations Manager in London and was then 'given this job as programme manager'. He describes it all as a fun career.

Inmarsat is an internationally-owned co-operative in the business of chucking satellites into the sky, building earth stations in remote corners of the planet and then using its assets to provide a wide range of communication services.

### Talking through space

There are currently a wide range of things you and I could do with Inmarsat's help and a great deal more just around the corner. Ship-to-shore telephones were originally important enough to name the organization: INternational MARitime SATellites.

When you watch live transmissions from remote corners of Bosnia, India or Africa do you stop to wonder how the images and sounds are being transmitted to your lounge? Do you ask why the commentators have so many pockets? Do you wonder why throughout the world children of all ages feel the need to get behind the commentator and wave? The answer to the first question is generally Inmarsat. The TV crews have a couple of suitcases loaded with camera, microphone, satellite dish and power supplies. Set the system up, link directly to the satellite and dial your editor and you've got Kate Adie live from the hills over Bosnia on breakfast TV.

The next generation of hardware will replace the luggage with one briefcase and then we should get phones in our pockets that search for the nearest link which might be the local cellnet system but, if you happened to be climbing Kilimanjaro, would probably be the nearest satellite. That'll put you one up on the guy in the Golf Gti on the King's Road with a Vodaphone stuck to his ear.

Guess who is behind the installation of phones in Jumbojets? Inmarsat Aero is installing flying handsets that talk to satellite and therefore to anywhere from within a plane. Air Traffic Control (ATC) could be revolutionised by these better links as a permanent electronic signal could give a better idea of where a plane is, who it is, where it is going and how long the in-flight movie has to run. Satellite-based ATC gives the potential of a single worldwide ATC system helping scheduled flights divert round weather at the cost of a phone call. Plus you could fax, phone, place bets, do your shopping and ring ahead to report your arrival at Los Angeles. Is there no escape?

Inmarsat and BT provide Whitbread round-the-world yacht race entrants a suitable transmitter. Those transmitters allow us to experience everything about rounding Cape Horn in a snow storm at 30 knots except the cold and the danger.

Inmarsat C – one of the newer systems – offers telex for small yachts and trucks. Take a driver on his way East as part of an aid convoy. The driver has a little terminal hooked up to an antenna on his roof. From time to time the driver taps in his location and it gets sent back to head office. They can make a message appear on his screen perhaps to reroute via a new pick up.

And finally it is Inmarsat who provide us with our daily diet of financial facts and figures. The TV stations buy a feed from Reuters detailing all those exciting market indicators and exchange rates. I know no one who is actually interested in the Nikkei, FT100 or the more friendly and American Dow Jones. In news-gathering terms it is satellite cheap.

**World wide**

There is a wide range of active projects in hand in the organization. There are new communication systems, new ground-control stations, new products galore. Satellite systems involve the space craft itself talking to and being controlled by ground stations. Ground stations are those remote sites with innumerable large dishes pointing up at space. The hardware is augmented by a wide range of software control and communication systems.

Rowe's projects cover a wide range and a wide geographical area. A small job is around $10m and large means $2.5 bn.

The current Inmarsat 3 ground control programme involves five major projects: Beijing (four antennas, two new and two upgrades); Fucino, Italy (four antennas – three upgrades and one new); Pennant Point, Nova Scotia, Canada (two new antennas) and Lake Cowichan, Vancouver Island (two new antennas).

The sites are chosen for remoteness. Deserted places tend to give less interference problems but cause significant problems when it comes to access. Nipping down the hardware store for a bag of nails in any of these locations is a few days, not a few minutes. Rowe spends a deal of time travelling between these remote locations and living in temporary site accommodation.

Back in London is one of those wonderful control rooms where banks of screens and operators look after the global system. The control computer system with hardware from Encore is based on RISC machines ($0.5m worth) and one project involves buying new and upgrading existing software. On my visit, we caught a satellite controller eating a Big Mac. It's nice to know they're human. (The controllers, not the burgers.)

The system allows a controller to type a command in London which is transmitted to China via cable or Intelsat (you don't use your own systems to control your own systems). This message gets processed through the baseband system, put into the RF system, transmitted to satellite. The space vessel might manoeuvre, or switch in a component. Telemetry sends back status messages all the time to let the ground-based controllers know what is happening above their heads.

I thought Keith might be able to give us a world wide view of management attitudes in these four very different environments.

'You have to be flexible, look at things, deal with people', he began, 'you might be successful in one place but you won't be in all four. There are different contractors and project management techniques, different contracts and attitudes.

'Canada is very USA-orientated', he continued, 'The American way of managing and dealing with things. Fucino is more

European, more laid back whilst still efficient. They are very Italian – a meeting starting within ten minutes of plan is OK to the Latin temperament.

'At the Chinese site few speak English therefore I need translation. They mostly tend to listen to you in English and then fire back in Chinese. They try to do everything internally to avoid importation. If there is a choice they'll do it in China even if it costs loads more. They learn very fast – we tell them we need improvements and they just do it. I don't underestimate the Chinese. They have a lot to learn but boy are they learning fast! The culture is still ancestral with an over-emphasised respect for age. This is starting to go a bit, but still there is great respect for wisdom. We decided that to have the right person leading the team was vital and to have older people working with young whizz kids. UK people have to have the right position and preferably be a bit older.

'We get special problems in these very different and remote locations. On the Lake Cowichan project we had a problem with the backing on some antennas which are made of foam. Woodpeckers pecked the foam to pieces. It had to be wrapped in steel but the manager still talks about hearing woodpeckers trying to get through on the transmission.

## Introducing project management

Historically Inmarsat didn't have much of a project-management process. Some projects were successful, some were not. There was no specific methodology. When a project failed it was normally due to poor initial planning. Keith concluded that 'we needed better initial processes to show that you can use a process and that process can be made to work and be successful'.

So they set up three or four projects to follow a more formal process of planning and management including the selection of personnel. One of these special projects was Keith Rowe's who explains: 'These projects were given every opportunity to succeed but had a microscope trained on them. The Director General looks at these project and each has an Executive Vice President.

When Rowe joined in 1986 Inmarsat employed 200 people. Today with 600 people the organization has changed a lot. He takes the view that 'as a company gets bigger you need better systems'.

On the four test projects each team developed their own systems and out of these will come corporate methodologies. As there is such a wide range of projects – some projects are hardware, others more collaborative – teams are encouraged to develop processes on the hoof to suit their environment. We look for commonality between the programmes and which differences we should acknowledge. There is a team of eight or nine project managers under a Director plus an organizational development person working on a mini-project to set up a programme management system'.

Rowe enthuses, 'We have a library of tools, a handbook of systems, development plans for training and assessing project managers. There is a feed back from the pilot projects to this work. We compare notes and select the ideas that work for us.

'This is better than buying PRINCE. PRINCE is one tool in our library alongside training courses, planning software, risk software, books and references.

Ground Control Project (GCP) was the first pilot scheme under Executive VP Gene Jilg whom Keith describes as 'a project champion, a powerhouse. He pushed through the need for professional project management plus moved the company from annual budgets to project budgets where appropriate'.

This is often a problem within organizations where the annual budgeting system conflicts with multi-year project life spans. Inmarsat now operate on appropriate projects a system with revised budgets each year for multi-year projects. GCP has an overall budget which is resubmitted each year. If it fits within the original cash flow envelop everything is OK. Keith reports, 'GCP is still on schedule and still on budget after moving two years down the road. This is due to good planning and budgeting. It is a credit to those who work on the project'.

Ah, comments your suspicious reporter, one way of being on time and to budget is to start with generous allowances for both. Keith's reply seems honest. 'Time was not generous but money was slightly. I think overall it balances out. We have not yet hit the contingency.

'We believe in setting a realistic sum and time plus a contingency and then to tell the team to go and do it. We don't bitch and moan'.

### The project team

Keith set up a programme office to deal with Inmarsat 3 ground system programme.

'We have eight in the programme office essentially managing contracts with nine in TTC doing the same. There are eleven involved in implementing the computer project plus eight in operations and four in flight dynamics. This gives us a team of about 40 virtually full time plus contractors and their staff all over the world working on a total project worth $350M'. This is some asset.

My impression of the people I met at the London offices of Inmarsat was the most unlikely conceivable for a group of people working in space: they have their feet firmly on the ground.

I suppose that every visitor to Inmarsat will come away with their heads slightly in the clouds. Perhaps even above the clouds. We tend not to think about what is in orbit or how those bits got there but we just enjoy the benefits without a second thought. Next time Aunt Marjorie calls from Australia, you watch a TV showing some remote part of the world or take a call from your stock broker in mid Atlantic blame those quiet, unassuming and very down-to-earth people at Inmarsat for having made all these things possible.

# Quality and BS EN ISO 9000

## Concepts

There are a good many misconceptions about this topic and a good question to begin with is: what is quality? Quality does not mean good, deluxe or excellent, it means meeting expectations.

If you are travelling through Asia as a seventeen-year-old student with a rucksack on your back, a song in your heart and freedom of mind in your stash box you have certain expectations of the place you will stay each night. At the top of the wish list is the company you will be keeping. Other like-minded travellers with whom you can chat are very important. The question of cost will arise the next time you pat your money belt to remind you that cheap is the best you can afford.

If you meet up with some other travellers and find a cheap little place right under the ancient city wall where everyone sits around and smokes and chats all night you will be dead pleased. The absence of hot water will be just one of those things and the presence of curious, many-legged, hard-shelled insects only adds spice to your stay. When your stay costs you the equivalent of a UK bus fare you will be delirious. It met your expectations.

Now step forward 40 years. You travel on business to Asia in your fifty-seventh year. Your idea of a good night now is a five-star luxury hotel with gold bath taps, hot and cold running servants and a nice restaurant. You might use the in-room phone, send a couple of faxes and watch satellite TV. The flunkies are dressed in something that looks a little like a traditional national costume. The dress might have been traditional somewhere but nowhere near the city in which you are staying.

If you got looked after like royalty and spoke to no one apart from ordering a meal and the odd 'thank you' you would feel like this place fitted your bill. The actual bill has no bearing on your feelings at all as it just goes on the personal expenses form. You got what you wanted. You would have hated the place 50 years ago and would only have

considered staying there if you were very ill and someone back home was picking up the bill. You would have felt out of place and hoped other teenage travellers didn't see you anywhere near the door.

Hence quality is all to do with expectations. Here comes a formula – this must be a proper book.

$$\text{Quality} = \frac{\text{Delivery}}{\text{Expectation}}$$

This little formula tries to say in a quasi-mathematical sort of way that if you take that which was delivered and divide it by that which was expected you get a value or measure for quality. If quality equals one then everyone should be happy, less than one means that someone is going to be disappointed. If the ratio is over one, someone should be more than a happy chappy.

It is not easy to measure any of these three concepts. Your local builder's merchant probably does not sell a happiness gauge. It is hard to find a yardstick with which to measure people's expectations. If you wish to try, nip along to your local airport whilst another air traffic controllers' strike is making life entertaining for mothers of three screaming kids waiting to go to Alicante and discuss with her the intricacies of measuring expectations and delivery. For that matter try the same thing with any woman during child birth – at least she has delivery on her mind.

These measurements are subjective, it is all about what people feel about quality. They can be disappointed, delighted and often a complex mixture of both. If it is a matter of opinion the big issue therefore becomes: whose opinion counts? The answer comes back as easily and as clearly: the client's.

If you are in the business of managing projects then it is your client's opinion that really counts. Whatever you do, how every much you try, in the end the measure of the quality of the project is held mostly subjectively within the client's mind.

This is one of the reasons that leads organizations that set about projects where there is no obvious client into creating one. Imagine you work for an engineering contractor and your project is to build and deliver a bridge to the government of Thailand. The client is obvious – you have to make sure that the Thai government's representatives are happy.

Now imagine your project is to organize the office move or introduce a new product. Whoops apocalypse, the client's disappeared in a puff of organizational change. Who is going to measure your delivery and expectation now?

You might say that this is easier. As there is no official client all you have to do is satisfy yourself and that is much easier to do. However, the chances are that there are a large number of people with fat cigars and bellies to match who are keenly interested in this new product you are managing. If they are not happy with the outcome of the project they, as board members, may well not think kindly of your efforts and this may have a deleterious effect on your future promotion prospects. In the absence of any other competition, your boss becomes the client.

When you don't have a client you are likely to:

1. Set your own expectations as you go.
2. Achieve an appropriate deliverable.

Own projects are a doddle. You may have moved your 'expectations' downwards a few times as your 'delivery' became clearer but you are likely to feel happy with the outcome whatever.

Many organizations deal with this problem by appointing a client whose job is to have expectations and to make them clear. It might still be the MD and his cigar may match his belly but you have then got someone else's opinion to worry about. The difference is now you know whose opinion to worry about. If you want to hit quality, first you must know who is setting the target and what they will regard as success and failure.

There is a slight variation on this idea and it goes like this – quality is about exceeding the client's expectations. This idea suggests that you should always try to please the client with a little more than he or she expected.

It makes total sense therefore to know your clients and what is in their minds. There are huge numbers of projects where the team worked their butts off and used all the best project-management techniques to achieve a wonderful project but misunderstood the client's needs. These mis-understandings set the quality formula to way below one.

Physical deliverables

A couple of points about the quality of a physical deliverable. Some industries have gone to enormous lengths to define quality in a measur-able way. In the engineering and construction industries there are all manner of technical tests available to decide if a piece of work should be acceptable. These are attempts to take the emotional aspects out of the picture and replace them with physical measurements.

Road builders and floor layers have little pyramid-shaped devices used to drop a standard steel ball a known distance onto the new surface to test hardness. Samples of concrete are taken from most building projects regularly and smashed to pieces at a laboratory to test their strength. If you drive past a housing estate you might spot small walls built in incongruous places. These are sample panels built to show the bricklayers what the walls should look like.

These are attempts to replace the quality formula with a specific mea-sure. Maybe you can do the same. It is much harder with non-physical deliverables.

Rework is bad for you

The down side of unacceptable quality in a deliverable is rework. When your project's deliverable fails to meet expectations it probably needs some repairs. This kind of reparation is generally called rework. IT people call it debugging. It is money that is spent purely and simply to raise the quality of the deliverable to approach the expectations of the client.

The new car, when test driven, rattles like a snake so it has to be dismantled and have its loose bolts tightened up. This cost is straight off the bottom line of your company. I attended a lecture given on this topic to a major defence organization whose rework costs were about £5m per annum. The lecturer pointed out that to increase profits by £5m would mean increasing turnover by £50m (assuming a 10% profit margin) which was nearly doubling their annual sales levels. He pointed out that reducing rework by 50% was like an extra £25m of profitable work –

worth spending some money to achieve. And the money that he recommended spending to achieve these savings was only a series of lectures, some posters and the spread of a quality ethos throughout the company. 'Getting it right first time, every time' is the phrase that springs to mind.

## The terms

**Total quality management (TQM)** is an approach to managing a company. TQM is a philosophy and I'll talk a little about these ideas later on. TQM helps you to achieve a quality product.

There are also a number of standards floating about. The old standards for quality, ISO 9000 and BS 5750, have been brought together into BS EN ISO 9000 which is a European standard. These are published books explaining the general implications of the standards. Organizations can apply for and get accreditation under the BS EN ISO 9000 scheme.

These standards are about how you do things, not what you do. If you want to produce the most appalling crap imaginable you could do so under a TQM environment and with accreditation. Just because the doubleglazing firm you are buying from has BS 5750 on their notepaper doesn't make their doubleglazing better, it just means their systems are good. They just happen to be making their leaky windows with great efficiency. They also need some new notepaper with the new European number on it.

If they don't show up to install your new windows on the right day, if they send the wrong components, if they send you an incorrect invoice you have every right to say 'I thought you were a BS 5750 company'. If the windows leak don't bother to bring the British or European standards into the conversation.

## Getting the badge and keeping it

I want to apologize to all those hardworking quality consultants who earn their living by helping companies achieve accreditation for calling this complex and expensive process 'getting a badge'. No, I don't really want to apologize at all – it is badge.

To be really on the ball you have to get the badge and then say in the press release something like 'We felt no need to improve our systems to achieve accreditation' just to show what a smart lot your company has always been.

To get the badge you have to show that your systems, procedures and practices are in good shape. You would find an independent quality auditor who will have been licensed to do this kind of thing by the National Accreditation Council for Certification Bodies (NACCB). It is even more likely that one of these quality auditors will find you and offer you their services.

Once you appoint a quality auditor he will begin by visiting your organization on a pre-audit visit to assess the state of your quality systems and suggest where these need to be improved. The quality auditor also programmes his own workload at your organization. Two or three months later comes the full audit with a full audit team which begins with a meeting to discuss what they plan to do.

The quality auditor will install a team of auditors and these will scurry all over your organization assessing the effectiveness of the practices you have in place. They will look at your stock-control systems, purchasing procedures, document-control mechanisms and a whole raft of similar areas of management.

They will not look at the doubleglazing manufacturing plant as they haven't the first idea about that kind of thing. The audit ends with a meeting with the senior management where problems preventing accreditation are discussed. Small problems get put right quickly, slightly larger issues are mentioned and you are given time to correct these. Serious issues are mentioned and you get time to fix these before a second audit.

It is all a bit like a management MOT. There is a curious relationship between the auditor and the auditee. You – as the company wishing to be accredited – pay the auditor's bill. It is a little like a defendant paying the judge's wages in the Old Bailey. But these are such professional people that mere money stays out of the equation until they start chasing you for payment.

When the auditor is satisfied, you get your certificate. It seems essential then to have a picture of the whole, happy, smiling staff taken with the MD holding the precious certificate and for this to form part of a press release which gets sent to a totally bored press who say 'another BSwhatsitsname'.

You get your notepaper and brochures reprinted with the certification logo and watch admiringly as very little happens. It is those who didn't bother who have to sit on the edge of their executive chair as their company misses out on all sorts of work because they are not certified.

It doesn't end there as the auditor may carry out spot checks twice a year for the next three years. Spot checks or no, three years later you are still not off the hook, but you must apply to be recertified if you want to keep your badge. The standards are not written with project management in mind and have to be bent and interpreted to the project-management situation so try to find an auditor who knows something about project management.

You might have the idea that I am not convinced by the BS EN ISO 9000 system. You're right, to my mind it is a great idea but it has become accepted as something it never pretended to be. As a project manager once said to me: 'We've been tightening nuts for years here, now we have write down how to do it.'

## Quality in the project-management environment

Apart from the accreditation route someone in your organization might decide that the company needs total quality management. The fact that he heard about TQM at the golf club and hasn't a clue what it means is immaterial. TQM in a nut shell is a quality approach applied to four things:

- the organization
- the management process
- the operating process
- the product

In a moment I'll explain something about the ideas that TQM organizations use to improve quality and reduce the costs of rework and wastage. But first a word from our sponsor.

A slight barrier is that quality assurance or quality systems are generally designed for the production environment. There are problems in the project environment which are different to those in the production environment and there are people who understand these differences. Some of those differences are:

- There can be problems having any kind of standard systems when the range of projects varies widely or when dealing with one-off or strange projects. The systems that get set up to help monitor the quality of baked beans as they zip down the production line are going to be totally inappropriate to the production of a document or the launch of a product if that is the one and only time you are likely to do it.
- You don't have much time to set up systems in the project life cycle. You get the go-ahead and it is 'all hands to the winch'. One of the things a good TQM system needs is time to set it up and time is something you are almost certainly going to be short of. If you are going to use TQM procedures you had better have them ready and available before the project gets underway.
- Who is going to pay for TQM support? You may not be willing to pay for TQM support out of your hard-earned project budget but are much more likely to suggest that these costs ought to be a company overhead.

If you decide you need help from a TQM consultant, choose one who can at least spell project management – they might understand these problems. If you're really lucky they might have some solutions. There are consultancies who specialize in TQM in the project-management environment.

So what is this TQM stuff? It is more than anything else an attitude. It is an attitude that you might try to create within your organization. Here are some of the pointers that are supposed to help engender this TQM attitude.

It starts at the top with a mission statement or a policy document that outlines the company's determination to be a 'quality organization'. Such statements are to be found enshrined in gothic script on an aluminium plaque screwed up in the reception area. Here is one for example:

> *This company strives to continuously improve its products, its relationships with its customers, its employees and the other organisations with which it deals.*

My bank has a more local and specific one:

> *We aim to serve everyone within two minutes.*

This means they will turn down your request for a bank loan with unprecedented speed. Behind this statement is a whole study of customers, response times, staffing procedures and all kinds of great stuff. Have you noticed how these days anyone on the bank's staff can march up and do something at any cashier's position? It took far too long when each position was looked after by one person who was responsible for the amount of cash in the drawer.

You'll have to imagine the aluminium plaque for yourself. Now, often this kind of statement really means what it says and the atmosphere throughout the company is accurately reflected in the words. Equally often, someone somewhere decided that the company needed some verbal pap to make it look good to all those well-meaning quangos and government departments which are systematically ripped off by the company's shoddy goods.

Your guess is as good as mine. A few days in the company will reveal where lies the truth. Some people think quality and some do not.

Some people spend their lives walking backwards looking at what has happened. They have a fairly clear view of what has happened and are able to complain accurately about the awful events that have disrupted their lives. Because they are not looking where they are going many awful events do happen to them which gives them much to witter on about. These people spend ages discussing how the ridiculous situation in which they now find themselves came about.

Others, and this should include all projects people, look firmly ahead and spend their time thinking and talking about what has to be done, what is going to happen. These people tend to make the same mistakes over and over as they rarely look back to see how well something actually went.

When something goes wrong these two types react very differently. By the words 'something goes wrong' I have in mind a failure of some kind – the software doesn't work, the missile casing is rejected or the satellite fails to reach orbit.

The 'backward-looker' spends the next few weeks in a desperate attempt to prove that it was nothing to do with him and it 'wasn't my fault'. His rear end is protected with a carbon-fibre reinforced deflector. If everyone agrees that it wasn't his fault, he might start to look at how repairs can be carried out at minimum cost.

The forward-looker immediately steps in with alternative solutions to resolve the problem, gets a plan prepared and, before you can say 'Henry Gantt had two Tees', has the repair works under way. Only when the repairs are in hand will he look at the cause of the problem 'so that we can avoid its recurrence'.

The backward-looker survives but may never end the project and the forward-looker finishes the project but goes miles over budget. Neither are perfectly right.

The TQM person would draw a clear line between putting things right and stopping things going wrong again. TQM says that you should have a policy of continuous improvement where you measure how well you are doing and then try to find ways to get better at doing it. The TQM idea is to get rid of the 'repairs department' by ensuring that it has nothing to do.

TQM aims squarely at the client and making sure that the client is happy. Typically a company adopting a TQM approach would issue a public statement pointing out the importance of the clients and the way in which the company was going to focus on them.

Perhaps a group of your employees go around your frequent clients and ask them questions about the service or goods you offer. What really pleases them, what distresses them? This gives many pointers as to what the company needs to do to improve its image. Go on then – when did someone from your organization last ask a client how well you did with a previous project? I have the idea that if restaurant owners did this sort of thing they might give you a few more chips instead of spending money on new flock wallpaper.

Some organizations go on to make each group within the company a 'client' and a 'supplier' all within the one organization. If you read the case study that follows you will see how each department that builds one part of a mainframe computer 'buys' the part-built machine from the previous department and 'sells' it on to the next. Therefore each department is a client who has to be satisfied with the goods it is getting and is also a supplier trying to satisfy someone else. Once this is all in place you need some measures of how well people, departments and divisions are doing.

Enter the graph shown here. All over the walls in TQM organizations you will see graphs comparing some measure of quality against time. It might be the number of failed components, it might be the number of complaints, it might be number of delays. I remember one company assembly line who graphed the failed components that they received by gluing the actual failed components to a large piece of graph paper in nice neat lines. The graph dealing with duff disks looked like this:

MAY JUNE JULY AUG SEPT

Not only could you quickly see the trend of component failures, you could see precisely which component the graph dealt with and you could even take a sample to see what caused the failure. This technique works well with microprocessors, electronic components but takes a little too much room if applied to offshore oil rigs or tunnels.

We're well on our way to TQM now, guys. We now need some **quality circles**. These are not nice, round chocolate roundabouts at the end of Quality Street. These are not even a highly ingenious, clever device which is hard to understand but very useful. It is a clever name for a very simple idea. It refers to a group of people talking together. It is a bit like a sewing circle except the topic of discussion is quality, not babies.

The quality circle notion has the organization forming small groups of people who are given time to discuss what quality means to them and their departments. They discuss how they are going to improve and what they are going to improve. They set themselves targets and try to achieve them.

This kind of thinking can go on in a project support office or even a quality support office. The danger is that quality in most people's eyes becomes a matter for 'those girls in the quality department' and 'It is their job isn't it?' No, it isn't. TQM is at best an all-pervading concept where everyone tries to improve and strives towards excellence.

A quality-support office can help, support and guide but it can't do quality for everyone else. If it tries it will quickly become a police department cracking down on quality crimes. So the best quality assurance (QA) people teach people how to look after quality but do not get involved with quality itself. The QA people guide and instruct but leave the actual quality issues up to the people on the shop floor.

Just a couple of more steps to quality illumination. How about writing down some of the better procedures. You could call the document produced by this process a quality manual. There are a number of advantages in this: you provide some useful employment for some people to study and record management and production processes; they might take a closer or more distant look at the procedures and suggest some improvements; and new people will be able to read about the way things are done. You might get BS EN ISO 9000 approval or be heading in that direction.

One disadvantage with all such documents is that they do not appear to generate additional income. This is almost entirely because they don't generate any additional income. Therefore, in many organizations, some young, inexperienced graduate or some more mature individual who is filling in time before getting a carriage clock and a cabbage patch is given the task of writing up the quality manual.

Having got someone inappropriate to prepare the manual, the company sets up no systems to modify or adapt it in the future. So the guys and gals that really know what they are doing have to blindly follow this outdated, inappropriate guideline feeling bad inside about the whole thing.

On the other hand, if some good people write down what they themselves do you will truly get advantages but spend some real money. The choice is yours. The next step is a quality system.

## Architecture of a quality system

It does make absolute sense to set up a quality system. This short section is a very brief example of a possible architecture for a quality system for a medium-sized research and development organization. It is included to show the sort of things you ought to be thinking about when and if you get into this field.

People seem to like neat diagrams with very few words so here is one to kick the system off:

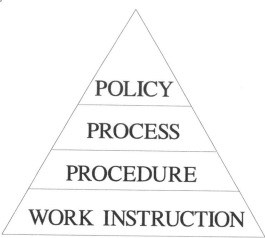

I like this diagram. I like the way that the designer has chosen terms that get longer lower down in the pyramid where is there is more room for them.

This shows the quality policy at the top of the pyramid as there is probably only one policy and this singular policy might be contained within a quality manual. This is a board or senior-management prepared statement describing the company's commitment to quality and the principles that the board wishes to be followed.

The process layer shows the processes by which the company manages its affairs. It is often diagrammatic or at least in part a picture showing data flow and other elements that define how the company goes about doing its thing. It might show who is responsible for the tasks that need to be done, who actually does them and is in the business of letting people know how the business functions. It might diagramatically show how a new project is tendered for. There are more processes than policies, but more procedures than processes.

The procedure layer details the steps or series of steps that need to be followed to achieve a specific process. This would, for example, show the steps that must be followed before a tender is submitted for a new project.

Next comes the work instruction. This is a level of even greater detail and work instructions provide additional detail to explain any step within a procedure. There will be more work instructions than procedures.

Running alongside the four levels of this pyramid would be standards. These provide a standardized or default specification of deliverables or outputs from any of the lower three layers of the pyramid.

An example of these layers that is close to my pocket and probably yours is the personal expense policy. If you detect a little surprise in your mind at this application you should read the case study at the end of this section which looks at a company that applied quality thinking to every aspect of the organization.

The quality policy document would probably make no specific reference to personal expenses but there probably is a policy document somewhere which lays down a policy for claiming expenses in very broad terms. This might well be included in your terms of employment. Below this there is almost certainly a process by which people claim and receive their personal expenses. This probably explains in general terms that there is a system, that people need authorization by a senior manager and that the purchase of alcohol is not permitted as an allowable expense.

Then there will be a procedure which explains that you need to get the chitties from the restaurants, hotels and other places to confirm your expenditure and fill out the appropriate form, get the appropriate approval and submit it to the appropriate section. It might say that you can draw an advance and how overseas expenses are dealt with. This might apply to everyone and therefore be a bit general.

The work instruction might apply to people at your level in your branch and in your situation. It tells you to get Form 134/GHB from head office, fill in a copy and send it to Room 101 in accounts. They will then send you a form 134/GHB/User on which you provide full details of your expenses attaching original copies of the invoices showing the VAT number clearly. This you get signed by the director responsible for your department and submit it to the wages department prior to the 24th of the month in which case you will get the reimbursement with your pay cheque in the month following.

This is close to the work instruction employed by an ex-employer of mine and it shows how such systems fail completely. My director lived and worked in another office so the local wages department didn't have a copy of his signature against which to check my expenses. They asked me to get a copy of his signature so I obliged. Being highly efficient I didn't trouble my busy boss with this trivial request but signed his name on a memo and sent it in to the wages dept. They happily checked my signing of my boss's name against my signing of my boss's name every month thereafter.

## Case study: Amdahl

This case study shows the way this advanced company approached quality and brought about a highly quality-orientated philosophy. It may be a few years old by the time you read it and things might well have changed within the company but it does show how a quality approach can be adopted very successfully. It has not previously been published.

This is the story of a typical Amdahl 5990 computer which has just been assembled by those very quality-conscious people who work in the Amdahl plant in Swords, Dublin.

It is a very fast computer which needs to be fed with copious amounts of electricity and is able to handle huge amounts of information. It is a quality computer. I know this to be a fact because the company is deeply committed to Quality. It even has a capital Q when they say its name.

The company and its products are named after Gene Amdahl, an ex-IBM employee who started making his own computers in 1980. He noticed that IBM had two chinks in its strategy. One was the huge profit percentage tacked onto their main frame computers and the other was that there was a graduated price structure across the range. Gene, a designer, built the ancestors of today's machines to compete with just one of the IBM machines and underpriced IBM. Not too much of a price cut, not too little. If IBM had felt like cutting Amdahl out of the market they had to drop their prices across the whole range of computers, something they decided not to do. So then we had two players in the main frame marketplace: IBM, Amdahl.

People have to choose one of these manufacturers and they talk about 'Blue' sites (IBM colours) and 'Red' Amdahl sites. I must not forget Fujitsu who have more recently joined in and who operate a technology exchange with Amdahl.

Gene's company shone amongst the many bright computer companies, made profits larger than most and prospered. More recently the general sales price of the main frame computer has started to drop so this company decided to get better and more effective.

Paul Renehan, Director of Product Assurance, said that: 'The company is committed to effectiveness not efficiency. It is very easy', he said, 'to do entirely the wrong things very efficiently, so effectiveness is a better objective'.

One of the key components in making the company into a 'World Class Company' is Quality and quality is about making sure that the client gets what he wants. What he does not want is to spend $7 million on a computer that doesn't even work and then not get help to get it mended.

So Paul Renehan and the others around the plant outside Dublin established a Quality Improvement Programme. The QIP has some very good ideas and concepts, so if you are interested in Quality, take note of what happens at Amdahl.

One of the theories goes like this. If we want to make sure the customer is happy, we must make sure every stage of the process, from order to installation, is done with the customer in mind.

Taking this to its logical conclusion they took each part of the process and called it a cell and made each cell the 'client' of the previous cell. For a partly built computer to pass along from cell to cell it has to be tested and accepted. Most of the computers expect to spend much more time being tested than assembled.

Let's take an example. The first stage of construction is the metal framework which is an equivalent of the human skeleton. The skeleton is accepted by the next cell (Printed Wiring Assembly).

The naked framework will sit in the assembly area – which is marked out on the floor. As this is a 'cell build' environment, it will sit in this square until ready to go on to the next cell. It won't get parked in a corner and it won't be passed on incomplete.

All the components are stacked around the cell in bins with the correct quantities in each bin. Each bin is marked and the items within it have already been tested.

The workers in this cell take components, check them, and mount them in the framework. If any component fails a test it gets marked up on a board with the date, time, nature of the problem all available for public gaze. Graphs around the building show how many faults were discovered over the last few months and the trend is reassuringly downwards.

They use a Just-in-Time approach to materials and components to ensure that the right parts are available at the right time.

When ready the part-assembled computer will be offered to and accepted by those nice people in the next cell: 'LSI and Array Production'. Once again the machine will sit obediently in its marked out rectangle as components are hung on and, more importantly, tested again and again.

Sometimes you think they just build things so that they can test them. The last few stages are all testing. There is: Diagnostic Testing, Margins Testing, Reliability Testing, Applications Testing and finally customer configuration. This testing process will take about six weeks.

Even the shipping is treated as a cell in the process of getting a computer to the customer. The computer will get crated up with a collection of meters that will monitor temperature, humidity and G forces on the journey to site. I heard about one machine whose records showed a gauge which had showed a blip in humidity and temperature during the trip. It turned out that the lorry driver had stopped to collect his motorbike en route. He had opened the back doors and put his bike in the back causing the blip on the dials.

The concept of quality runs right through the firm. The personnel department monitor how long it takes to respond to job applications and how long it takes to recruit each member of staff. Even in the light, airy, canteen there is a graph of complaints.

You might think that it was hard to bring in such a quality-conscious attitude. It all began back in July 1985 when the company met to discuss quality. The idea was to align motivations so that everyone worked in the same direction and to introduce a new attitude by evolution, not revolution.

Their first step was to train the senior management. They rented an external training school (feeling that an external school emphasised the importance) and trained 320 people under a Quality Education Scheme. This got the concepts over and left people feeling that they needed tools.

The next step was to form Corrective Action Teams (CATs) each of which looked at their customers' needs, requirements and defects problems. Don't forget that their customer would normally be the next cell in the production stage.

Each CAT chose a problem and worked at it. Perhaps they chose defects. They would collect and analyse data, design an action plan, implement that plan and monitor its performance.

Every section had to select three main areas and display their quality records for all to see. The senior management were delighted that these teams set much higher targets for themselves than the senior managers would have dared to set. The CATs tended to achieve the targets partly because they had set them themselves.

Quality improved so much that charts' scales had to be changed. Originally a chart might have showed defects as a percentage, but soon the graph had to be changed to defects per thousand so that the line could be seen.

Each CAT reports to a steering committee which co-ordinates developments from team to team.

The Amdahl team regard quality as almost free. I can understand that generally increased quality leads to less waste and a cost benefit, especially in a high technology environment like Amdahl.

I have also noticed two things about the people here. They are generally quite happy. They seem to be happiest when their management listen to them through the forums where anyone can contribute something to the way the company is run. People like to be respected and feel that they can change things for the better. There is a very good atmosphere.

The other thing I have noticed occurs when a VIP tours the plant. It could be the senior management or a client. I was one of an anonymous small group of people walking about in white antistatic coats.

When such visits are going on no one stops what they are doing, If they are working, they go on working. If they are having a chat, they go on chatting. In most factories most people feel they should 'Look busy, here comes the boss'. People at Amdahl are much more honest, straightforward and perceptive than that.

## *Good luck with your projects*

All that remains for me is to wish for you, dear reader, that which I wish for myself.

I wish you successful projects and, more importantly, a successful career in project and programme management.

Good luck with your projects – you'll probably need it.

# Appendix 1
# Automated assists

This is a list of software vendors in the programme-management marketplace. This is likely to be contentious as some organizations will complain that their product offers 'powerful programme-management features' and therefore should be included.

| Company | Address | Telephone and fax numbers | Products |
|---|---|---|---|
| ABTI | 95 Wandsworth Rd, London SW8 2HG | 0171 735 0800 | PMW, PMD |
| Lucas Management Systems | 219 Bath Road, Slough SL1 4AA | 01753 606060 | Artemis, Prestige Schedule Publisher |
| Computer Associates | 183 Bath Rd, Slough SL1 4AA | 01753 577733 | SuperProject |
| Primavera | Elsior House, 77 Fulham Palace Road, London W6 8JA | 0181 748 7300 0181 748 2846 | P3, Expedition |
| Welcome Software International | The South Bank Technopark, 90 London Road, London SE1 | 0171 401 2626 | Openplan Texim Timesheet |
| PDSI | Woking Eight, Forsyth Road, Woking GU21 5SB | 01483 727000 01483 727979 | PX, Maximo |
| Mantix Systems | Mantix House, London Road, Bracknell RG11 2XH | 01344 301515 01344 301083 | Cascade |
| Microsoft | Wharfdale Road, Winnersh Triangle, Wokingham, Berks RG11 5TP | 01734 270001 01734 270002 | MS Project |

| | | | |
|---|---|---|---|
| Digital Tools | Willem Parkweg 52-1<br>1071 HG, Amsterdam | 010 31 206 797751 | Autoplan |
| PSC | Bluegates, Willow Ave.,<br>Oxford Rd, Uxbridge<br>UB9 4AF | 01895 271272<br>01895 272677 | Max |
| Cosar | 6 Lanark Sq.,<br>Glengall Bridge,<br>Limeharbour, Docklands,<br>London E14 9RE | 0171 515 9700 | Trakstar |
| Leach<br>Management<br>Systems | 6–7 The Causeway,<br>Chippenham,<br>Wilts SN15 3BT | 01249 443118<br>01249 659895 | CS Project |
| Aran | Rivermead, Pipers Way,<br>Thatcham,<br>Berks RG13 4EP | 01635 872122<br>01635 871884 | PMS Kernel |
| Software &<br>Systems<br>International<br>Limited | 3 Bristol Way,<br>Slough SL1 3QE | 01753 528725<br>01753 694747 | Multi-project |
| Management<br>Systems<br>Consultants | 24 Cherry Orton Road,<br>Orton Waterville,<br>Peterborough PE2 0EF | 01733 391292<br>01733 51922 | MPMS |
| Symantec<br>Northern<br>Europe | Sygnus Court, Market St,<br>Maidenhead SL6 8AD | 01628 592222<br>01628 592393 | Timeline |
| Panorama | Walton Court,<br>Station Ave.,<br>Walton on Thames<br>KT12 1SH | 0181 891 0202<br>0181 891 6083 | Panorama |
| Tekware/<br>Scitor Corp | The Barclay Centre,<br>Worcester Road,<br>Hagley DY9 0NW | 01562 882125 | Project Scheduler 6 |
| Innate<br>Management | 419 Richmond Road,<br>Twickenham TW1 2EX | 0181 892 3637<br>0181 891 5027 | Innate |
| The Project<br>Group | 1 Mulgrave Chambers<br>26–28 Mulgrave Road<br>Sutton<br>SN2 6LE | 0181 770 9393<br>0181 770 9555 | Intime |
| SSI | Faraday Road,<br>Daneshill West,<br>Basingstoke RG24 8LH | 01256 51821<br>01256 858238 | Protos 2000 |

# Appendix 2
# Contacts for more information

This is an easier one – it is a list of people and organizations to contact for more information about programme management.

ProgM – The Association of Project Managers Specific Interest Group in Programme Management

| | | |
|---|---|---|
| Chairman: Geoff Reiss | 01943 466025 | 01943 466025 |
| Secretary: Dean Hutchinson | 0171 637 9111 | 01734 326100 |
| | | |
| Association of Project Managers: | 01494 440090 | 01494 528937 |
| CCTA | 01603 704719 | 01603 704817 |
| Cranfield University | 01234 754410 | 01234 751806 |
| Henley Management College | 01491 571454 | 01491 410989 |
| 'Project' editorial office | 01491 652116 | 01491 652564 |
| Project Manager Today | 01734 761339 | 01734 761944 |

# Further reading

Abdel-Hamid, T.K. (1993) A multi-project perspective of single project dynamics. *Journal of Systems & Software*, September 1993.

Bock, D.B. and Patterson, J.H. (1990) A comparison of due date setting, resource assignment and job pre-emption heuristics for the multi-project. *Decision Sciences*, **21**.

Bock, D.B. and Patterson, J.H. (1990) A comparison of due date setting, resource assignment and job pre-emption heuristics of the multiproject scheduling problem. *Decision Sciences*, Spring 1990.

CCTA (1993a) *An Introduction to Programme Management*, HMSO, ISBN: 0-11-330611-3.

CCTA (1993b) *Managing Programmes of Large Scale Change*, HMSO, leaflet.

CCTA (1994) *Guide to Programme Management*, ISBN: 0-11-330600-8.

Coulter, C. (1990) Multi-project management and control. *Cost Engineering*, October 1990.

Dean, B.V., Denzler, D.R. and Watkins, J.J. (1992) Multi-project staff schedules with variable resource constraints. *IEEE Transaction on Engineering Management*, **39**.

Deckro, R.F., Winkofsky, E.P., Hebert, J.E. and Gagnon, R. (1991) A decomposition approach to multi-project scheduling. *European Journal of Operational Research*, 6 March, 1991.

Dumond, E.J. and Dumond, J. (1993) An examination of resourcing for the multi-resource problem. *International Journal of Operations and Production Management*, **13**.

Ferns, D.C. (1991) Developments in programme management. *International Journal of Project Management*, August 1991.

Gareis, R. (1991) Management by projects: The management strategy of the 'new' project-orientated company. *International Journal of Project Management*, May 1991.

Harrison, J. (1993) BP's culture change programme. *Training & Development*, December 1993.

Heck, M. (1993a) High end project managers. *Infoworld*, 1 February, 1993.

Heck, M. (1993b) Project manager bridges high and low end. *Infoworld*, 12 July, 1993.

Kurstedt, H.A., Gardner, E.J. and Hindman, T.B. (1991) Design and use of a flat structure in a multi-project research organisation. *International Journal of Project Management*, November 1991.

Leachman, R.C., Dincerler, A. and Sooyoung, K. (1990) Resource constrained scheduling of projects with variable intensity activities. *IIE Transactions*, **22**.

Levene, R., Gove, D., Watton, R., Potts, S. and Trotman, M.A. (1991) Multi-project environment: Strategies and solutions. *Project Manager Today*, 22 October, 1991.

Levene, R., McFarlane, D. and Koppleman, J.M. (1992) Multi-project management: Organisational problems and strategies. *Project Manager Today*, 10 March, 1992.

Levine, H.A. (1991) Projects goals shifting toward enterprise goals. *Software Magazine*, December 1991.

Lonergan, K. (1994) Programme management. *Project – the Journal of the Association of Project Managers*, July 1994.

Lonergan, K. and Dixon, M. (1994) Managing control systems in a programme environment, *Project – the Journal of the Association of Project Managers*, October 1994.

McCormick, E.H., Pratt, D.L., Haunschild, K.B. and Hegdal, J.S. (1992) Staffing up for a major programme. *Civil Engineering*, January 1992.

Marsh, D. *et al.* (1994) Programme management & project support offices. *Unicom Seminar*, 4 October, 1994.

Marsh, D. *et al.* (1994) Programme management & project support offices. *Unicom Seminar*, 17 May, 1994.

Nkasu, M.M. (1994) COMSARS: a computer-sequencing approach to multi-resource constrained scheduling. *International Journal of Project Management*, August 1994.

Nunamaker, J.F., Jr (1993) Automating the flow: groupware goes to work. *Corporate Computing*, March 1993.

Palmer, B. (1994) Programme management in the public sector. *Project – the Journal of the Association of Project Managers*, September 1994.

Palmer, R. (1995) Practical programme management. *Project Manager Today*, January 1995.

Palmer, B., Reiss, G. and Gilkes, P. (1994) Proceedings of the programme management seminars, *Project Manager Today*, 23 March, 1994.

Panday, K. (1994) No consensus on how to meet deadlines. *Management Consultancy*, July/August 1994.

Payne, J.H. (1993) Introducing formal project management into a traditional, functionally structured organisation. *International Journal of Project Management*, November 1993.

Platje, A. and Seidel, H. (1993) Breakthrough in multiproject management: how to escape the vicious circle of planning and control.

*International Journal of Project Management*, November 1993.

Platje, A., Seidel, H. and Wadman, S. (1994) Project and portfolio planning cycle: project based management for the multi-project challenge. *International Journal of Project Management*, May 1994.

Reid, A. *et al.* (1994) Implementing and operating project teams. *IBC Technical Services seminar*, 24 November, 1994.

Reiss, G. (1994) Programme management, parts 1 and 2. *Project Manager Today*, May and June 1994.

Reiss, G., Marsh, D., Burton, D. and Curtis, P. (1994) The role of the programme and project office. *Project Manager Today*, 8 November, 1994.

Scheinberg, M.V. (1992) Planning a portfolio of projects. *International Journal of Project Management*, June 1992.

Sipos, A. (1990) Multiproject scheduling. *Cost Engineering*, November 1990.

Talentino, J. and Sletten, J. (1990) Closing the gap: A multi-project management challenge. *Cost Engineering*, November 1990.

Tavener, I., Hurley, D. and Innes, D. (1993) Resources versus projects. *Project Manager Today*, 9 November, 1993.

Tsubakitani, S. and Dekro, R.F. (1990) A heuristic for multi-project scheduling with limited resources in the housing industry. *European Journal of Operational Research*, 6 November, 1990.

Turner, R., Reiss, G., Watts, M. and Meade, R. (1994) Programme management: The software challenge. *Project Manager Today*, 22 March, 1994.

Turner, J.R. and Speiser, A. (1992) Programme management and its information requirements. *International Journal of Project Management*, November 1992.

Watto, R., Newman, P. and Brown, J. (1992) Multi-project management: Tools & techniques, *Project Manager Today*, 10 March, 1992.

Whitson, B.A. (1992) Managing many projects the easy way. *Facilities Design & Management*, June 1992.

Williams, T.M. (1993) Effective project management in a matrix-management environment, *International Journal of Project Management*, February 1993.

Wirth, I. (1992) Project management education: Current issues and future trends. *International Journal of Project Management*, February 1992.

Wirth, I. and Hibshoosh, A. (1988) Decision support system for multi-project scheduling of resources in blood collection programmes. *International Journal of Project Management*, November 1988.

# *Index*

# Index

*International Journal of Project Management*, November 1993.

Platje, A., Seidel, H. and Wadman, S. (1994) Project and portfolio planning cycle: project based management for the multi-project challenge. *International Journal of Project Management*, May 1994.

Reid, A. *et al.* (1994) Implementing and operating project teams. *IBC Technical Services seminar*, 24 November, 1994.

Reiss, G. (1994) Programme management, parts 1 and 2. *Project Manager Today*, May and June 1994.

Reiss, G., Marsh, D., Burton, D. and Curtis, P. (1994) The role of the programme and project office. *Project Manager Today*, 8 November, 1994.

Scheinberg, M.V. (1992) Planning a portfolio of projects. *International Journal of Project Management*, June 1992.

Sipos, A. (1990) Multiproject scheduling. *Cost Engineering*, November 1990.

Talentino, J. and Sletten, J. (1990) Closing the gap: A multi-project management challenge. *Cost Engineering*, November 1990.

Tavener, I., Hurley, D. and Innes, D. (1993) Resources versus projects. *Project Manager Today*, 9 November, 1993.

Tsubakitani, S. and Dekro, R.F. (1990) A heuristic for multi-project scheduling with limited resources in the housing industry. *European Journal of Operational Research*, 6 November, 1990.

Turner, R., Reiss, G., Watts, M. and Meade, R. (1994) Programme management: The software challenge. *Project Manager Today*, 22 March, 1994.

Turner, J.R. and Speiser, A. (1992) Programme management and its information requirements. *International Journal of Project Management*, November 1992.

Watto, R., Newman, P. and Brown, J. (1992) Multi-project management: Tools & techniques, *Project Manager Today*, 10 March, 1992.

Whitson, B.A. (1992) Managing many projects the easy way. *Facilities Design & Management*, June 1992.

Williams, T.M. (1993) Effective project management in a matrix-management environment, *International Journal of Project Management*, February 1993.

Wirth, I. (1992) Project management education: Current issues and future trends. *International Journal of Project Management*, February 1992.

Wirth, I. and Hibshoosh, A. (1988) Decision support system for multi-project scheduling of resources in blood collection programmes. *International Journal of Project Management*, November 1988.